Also by Gerald Summers

THE LURE OF THE FALCON

AN AFRICAN BESTIARY

Gerald Summers

ILLUSTRATED BY
Juan Carlos Barberis

SIMON AND SCHUSTER
NEW YORK

Copyright © 1974 by Gerald Summers
All rights reserved
including the right of reproduction
in whole or in part in any form
Published by Simon and Schuster
Rockefeller Center, 630 Fifth Avenue
New York, New York 10020

Designed by Irving Perkins
Manufactured in the United States of America
by American Book-Stratford Press, Inc.

1 2 3 4 5 6 7 8 9 10

Library of Congress Cataloging in Publication Data

Summers, Gerald.
    An African bestiary.

    1. Zoology—Africa. 2. Zoologists—Correspondence,
reminiscences, etc. I. Title.
        QL336.S93   828.9'14'07   74-6127
            ISBN 0-671-21813-1

This book is dedicated to all the birds and beasts who appear in it and above all to the memories of Cressida the kestrel, Bracken the lurcher, and Rupert, my African pi dog, but for whom it is doubtful if a word would ever have been written, and who, by their close and loving companionship, inspired me to keep going at times when things were at their blackest.

CONTENTS

	Foreword	9
1	Journey to Africa	13
2	The Tanganyikan Farmer	24
3	A Pup Named Rupert	39
4	Shark Hunting on Lamu	57
5	Four Become Three	74
6	A Double Loss	87
7	Audax Joins Dingaan	106
8	Locust Hunter-at-Large	121
9	The Rhino and the Baboon	135
10	A Home for Karen	155
11	Mau Mau Menace	173
12	A Proud Saluki	189
13	Kwaheri Kenya	208

FOREWORD

I SAT in a wickerwork chair in the palm-shaded courtyard of the New Africa Hotel in Dar-es-Salaam, in May 1947, idly watching a pair of iridescent plumaged sunbirds—the hummingbirds of Africa, which has no true hummers of its own—busy about the scarlet-flowered poinsettias, while a tiny, almost transparent, gecko lizard, flat against the stem of a nearby borassus palm, watched me from great bulbous eyes that seemed much too big for its flattened, narrow head that looked like that of a toy crocodile. This was the Africa that I had always imagined and I felt a glow of contentment tinged with anticipation stealing over me. I glanced at my two companions. Cressida, the kestrel, who had been my companion in peace and war for five long years, sat on the back of my chair, paying no attention to me. She was watching the gecko, her head bobbing in true falcon style as she weighed her chances of launching a sudden attack on the unsuspecting reptile. On the rim of a large stone vase, filled with some spiky, unattractive cactus-like plant, perched young Tawny, the owl, which I had rescued from shabby captivity in far distant Petersfield in Hampshire. He, too, appeared deep in his own thoughts, his lustrous blue-black eyes half shut against the still strong evening light—blissfully unaware that he was probably the sole representative of his species in the entire African continent south of the Mediterranean. It seemed months rather than weeks since the three of us had boarded the troopship *Empire Ken* at Southampton on a blustery, typically English April morning to set sail for Africa and a new life, never, so I thought at the time, to return. As I sat with a much-needed glass of brandy at my elbow I thought of all that had led up to this and particularly of the long, eventful sea voyage, which had at last come to an end. Above all, I thought of Bracken, the shaggy-coated lurcher bitch, and the constant companion of my young manhood. After much soul-searching, fearing that her advanced age, the uncertainties of travel and the intense heat of Africa would be too much for her, I had decided to leave her at least temporarily in my mother's care, hoping that she might join me later, when I had made some sort of a life and home for her.

It was back in April 1945 that Cressida and I, liberated from our prison camp by the onsweeping Allied armies, had returned to England. It was to be nine months before I was demobilized, a time of increasing boredom and frustration. Cressida, however, seemed quite unconcerned and indeed had become something of a national heroine. We were photographed together, appeared on television, and I gave a talk about her on "Children's Hour." She took her newfound fame with her usual equanimity, and only on one occasion did she forget herself and indeed nearly came to a sticky end. She had flown at a sparrow in a busy street in Newcastle and a double-decker bus rolled right over her, but she emerged defiant and unscathed! After demobilization I spent some months traveling around, accompanied by Cressida and my lurcher, Bracken, visiting friends and generally enjoying myself in an uneventful, peaceful manner. This was all very well, but my family pointed out to me that I must soon make up my mind what I wanted to do; I should have to be independent and earn my own living. I had often had vague ideas of a career in show business, but as I was completely lacking in any experience and my heart wasn't wholly in it, it seemed sensible to look elsewhere. Indeed my only qualification, if it could be so called, was a very real love of and understanding for anything in any way connected with the animal kingdom. What chance, I thought despondently, had I of finding any kind of congenial employment?

My fears proved unfounded when I met one Colonel Penn, the London agent for Lord Chesham, who was then running a scheme for settling ex-servicemen on farms in the Southern Highlands of Tanganyika. The description he gave of life out there tempted me to catch the first available ship bound for this paradise, but I was practically reduced to camping in Leadenhall Street before I could get a passage. As a small boy I had heard much about Africa as a country full of wild beasts and wilder people. I had read more, and I knew even then that I must and would spent part of my life there. The fact that the country had cost my father his life in no way diminished my intense desire to pay it another, longer and more extensive visit.

1

Journey to Africa

IN THE IRON-HARD winter months before I left London, my mother took a house in Chester Row, and here Bracken, Cressida and I spent much time. Cressida would sit happily on her block in the little back garden with her crop stuffed with a variety of strange meats, for rationing was still stringent. Ordinary meat was almost unobtainable, but I knew a number of friendly and accommodating game dealers who kept us supplied with rooks and jackdaws, masquerading as black plovers, and even such exotic items as woodcock and partridge. The stand-by was the common feral, or street pigeon, purloined I suspect from the vicinity of Trafalgar Square. Cressida thrived and waxed fat and glossy on this diet. We used to brave the bitter weather to feed the clamorous hordes of black-headed gulls, moorhens and ducks on the frozen lakes in the various parks in central London.

Once I was asked by the actress, Nancy Price, to give a talk at the hall in Russell Square where she used to hold literary discussions and poetry readings. I went with Cressida and it seemed to be a success though I must confess to having been paralyzed with terror at my first essay at public speaking. I came to know Nancy Price well in the weeks that followed, and was a regular visitor to her flat where I made friends with her Amazon parrot, Boney, a bird of uncertain temper and advanced age. Boney at first eyed Cressida with the distaste that only a parrot can express, and he didn't think much of me either but, though he never unbent toward Cressida, I was eventually allowed to scratch his poll, which I suppose was a point gained. Nancy Price was interested in animal and, more particularly, bird welfare, and we used to have many discussions on this subject. She was especially fond of owls, and once appeared in the film *The Three Weird Sisters* with my tawny owl, although the latter did not take a very prominent or memorable part.

I had discovered the owl in the backyard of a down-at-heel, retired circus owner and animal dealer. I was anxious to acquire a mate for Cressida and had gone there in the belief that the man who owned this ramshackle collection of assorted unfortunates had a male kestrel, the idea of founding my own strain of semidomesticated kestrels being at that time strong. As it turned out he had none, but only a number of the more conventional wild beasts of the type usually associated with third-rate traveling menageries. There was a group of dejected-looking dingoes and a few half-grown lion cubs, a llama or two and an unhappy dromedary, all of which I was thankful to hear were being disposed of within the next few days to one of the better-known zoological gardens. I was about to leave when a decrepit rabbit hutch, half covered with a damp sack, caught my attention. Peering in, I saw a pair of huge lambent blue-black eyes set in a sad monkey-like face gazing up at me. Nothing can look more woebegone than an unhappy owl. I bullied and pleaded with the fellow, threatening him with the R.S.P.C.A., the R.S.P.B., and every other such organization I could think of, and finally after the rapid exchange of a ten-shilling note I left with young Tawny confined in a bicycle basket.

Back at home I examined him in some detail. He was a young male of the handsome rufous phase not uncommon among British tawny owls. Both his wings had been clipped and his tail was worn down to a stump, but he was, surprisingly, plump and remarkably tame despite his recent predicament. His appetite was excellent and he immediately wolfed down a large plateful of chopped-up raw horsemeat and liver. I introduced Tawny to Cressida, who went for him without a moment's hesitation. The owl was heavier and better armed, but the kestrel was the quicker and more aggressive of the two, and in a matter of seconds each had locked talons with the other. They lay on their sides gripping and struggling spasmodically; indeed I had quite a job to separate them. They soon learned to tolerate each other, however, and remained in a state of armed neutrality from then on.

At last, in April, came the news I had been so eagerly awaiting; a berth on the troopship *Empire Ken* was available, and the ship was leaving for Africa almost immediately. I quickly established that no livestock was permitted to travel with her passengers. This problem would have to be overcome by subterfuge, for it was unthinkable that either of the birds should be left behind. But my impending departure was not wholly a matter for rejoicing, for it meant perhaps a final parting with the best friend I had ever had. Only those who have loved and been loved by a good dog, who have had years of shared experience, sad and joyful, will know how I felt. My mother had bought a house near Battle, in Sussex, and the last few days before I sailed were spent there. It was springtime and Bracken and I walked in the woods alone together. The woodland floor was white, with drift upon drift of wood anemones, with here and there the upward thrusting head of an early purple orchid. I sat on a moss-festooned stump, and Bracken pushed her long bearded muzzle into my hand and looked at me with her wonderful dark eyes. I stroked her ears and rumpled her graying tousled head, and heard again the tiny gentle waterfall of song from a willow warbler, newly arrived from Africa, just as I had heard it in those woods near Brunswick on that April day in 1945, when Cressida and I had been released from captivity. But my thoughts were very different from those earlier ones, when it had seemed that I could never be sad or worried again. There came a rustle from a ferny bank across the ride; Bracken cleared the open glade in two bounds, and disappeared into the undergrowth, while I, my depression forgotten, hurried in pursuit. We never saw that rabbit, but I owe it a debt of gratitude.

On the day the *Empire Ken* was to sail from Southampton, I got up at first light and went for a last walk with Bracken around the chilly, deserted streets and squares of Belgravia. My sadness at leaving home was counterbalanced by excitement at the start of the great adventure. I was off to Africa at last! My mother came with me in a taxi to Waterloo station; we had always been close companions. We found my seat in the boat train and, after the usual platitudes common to the stiff-upper-lipped English on such occasions, the whistle sounded and the train began to lurch and jolt forward. I waved to my mother until I was nearly decapitated by a girder and then was suddenly alone; not quite alone, however, for on my knee I held my most precious piece of luggage. It was a brand new picnic basket, divided into two compartments; in one snuggled Cressida, the seasoned campaigner, and in the other sat Tawny the owl, about to start on the first long journey of his young life. Attached to the inside of the lid of this basket and so arranged as to deceive the prying eye of authority were two copies of *Lilliput*, and one each of *Men Only* and *The Argosy*, my pleasure reading for the voyage.

Journey to Africa

At Southampton we were ushered into an inhospitable and overcrowded shed where, after an interminable and nerve-racking wait, I duly presented myself with my suitcase, passport and basket before the Customs and Immigration officials. I noticed an opening at the far end of the shed that framed the end of the ship's gangway, up which some lucky passengers, their ordeal over, were already streaming. One of the officers looked at me, glanced at my suitcase and peered hard at my basket. He asked for my passport, which he scrutinized for an unconscionable length of time before returning it without comment. I gathered up my suitcase and, shifting my grip on the basket, started for the "gateway to freedom." "I suppose, sir," he asked suddenly, "you have not got a dog or cat in that basket?" "Certainly not!" I answered truthfully, striding up the gangway, half expecting to hear official footsteps hurrying after me. I found my cabin, which I was to share with three other cheerful expatriates, and stowing my suitcase and basket beneath my bunk, lay down to ponder my next move. The first and worst hurdle had been overcome, now for step number two.

One of my cabin mates was a young Scottish engineer bound for Nairobi, and Jock and I soon were on terms friendly enough for me to confide in him and seek his support. As we walked the deck together eyeing the other passengers and awaiting the first shattering blast from the ship's siren, I told him about Cressida and Tawny. He was enchanted and swore he would help me as much as possible. I now had at least one ally. As soon as the other two occupants came up on deck to explore the ship we returned to the cabin. I extracted the two birds from their incarceration and introduced them to Jock who, without hesitation, took Cressida upon his ungloved fist and began stroking her breast and tickling her nape. He admitted that he had never previously handled a bird larger than a budgerigar, but he soon fell for the charm of his two non-paying fellow passengers. When our other two cabin mates returned from their explorations we let them into the secret. As a result, Cressida and Tawny, within an hour of boarding the *Empire Ken*, had four staunch supporters behind them in case of trouble.

That night Jock and I drank deeply and returned to our cabin filled with a sense of well-being, by no means entirely the result of the duty-free brandy which, owing to the scarcity of Jock's national beverage, we had been forced to consume. Before climbing into our bunks we spread newspapers on the floor and gave Tawny and Cressida an hour or so of exercise, flying from bunk to bunk and scuttling about on the floor. I had brought a good supply of fresh horsemeat with me and both birds stuffed themselves before being returned to their traveling quarters. Tomorrow we would make better arrangements for them.

The next morning, despite a more or less sleepless night, I still felt reasonably optimistic as to what the day would bring forth. After an excellent breakfast, followed at no great interval by a very large brandy (for the ship was beginning to get a bit skittish as she approached the Bay of Biscay), I decided to risk all and deliver myself into the hands of my enemies, if indeed they should prove to be so. With Cressida on my fist and my heart in my mouth, I emerged from the cabin, escorted by Jock looking rather like an anxious but determined Scottish terrier. We climbed to the promenade deck and were immediately surrounded by passengers, who all showed a friendly interest in Cressida. A number of them had read of her adventures in the national press and were delighted and amazed to meet her so far from her native shores. Our triumphant progress was soon barred by the approach of Authority. A ship's officer appeared; he was polite, even friendly, but he asked me my name and the number of my cabin before disappearing up a companionway, obviously to make a report. Within an hour the blow struck; over the ship's tannoy came a message, deep and sonorous: "Would Mr. Summers report immediately to the Captain on the bridge?" Wondering whether I should be given the option of keelhauling, walking the plank, or merely hanging from the yardarm I had no alternative but to obey. After a reassuring clap on the back from Jock I slunk in the direction of the bridge, which I eventually reached after a complicated journey along what appeared to be a nautical version of the Hampton Court Maze.

On the bridge stood Captain Jones, who at that moment looked more like Captain Hook. All he needed was a three-cornered hat and a cutlass between his teeth. He made no attempt to hide the fact that he was displeased, wasting little time on social niceties. "What's all this about you bringing livestock aboard my ship?" he snapped. "Is it true?" "Yes," I replied for want of anything more original to say. "Well, what the devil do you mean by it?" he thundered. Having always believed that a soft answer turneth away wrath, particularly when I was outranked, outgunned and generally out on a limb, I decided that frankness and what I hoped would pass for boyish charm might yet carry the day. I told him everything and was finally relieved to see his granite features relax a trifle. "Well," he said, when I had finished, "I suppose you realize that I could have them thrown overboard if I wished." "Oh, I am sure you wouldn't do that, sir," I replied. "Well, let's have a look at them anyway," he said. So I brought them to the bridge where, as I had hoped, they proved their own best emissaries and hurdle number two had been cleared. My stowaways were stowaways no longer. They were in fact probably the most popular of all the passengers and did much to relieve the monotony of the voyage. They used to sit with me out on the deck and take their baths in a canvas bucket, always surrounded by an admiring host of well-wishers. I am sure that the children especially will always remember that voyage and their unusual shipmates.

The commanding troop officer was an enthusiastic ornithologist and used to announce the arrival on board of any rare or unusual bird that happened to alight on the ship on its northward flight, for the spring migration was at its height. One of the migrants that visited the ship was a tree pipit, which arrived in an exhausted condition and lived for some time in a cage in the C.O.'s cabin, where it did very well on a diet of cockroaches of which the ship had more than its fair share. Eventually it was released at Port Said to begin its journey all over again. We also had a number of turtledoves and yellow wagtails, a nightjar and a hoopoe, and a few rarities such as melodious and olivaceous warblers. The C.O. would appropriate the loud hailer, and the whole ship would reverberate to the announcement "Information Birds," and this would be followed by particulars of the new arrival. This I regret to say caused not a little mockery among the more frivolous-minded passengers, but I was of course delighted to take advantage of any sightings that had escaped my notice.

To a naturalist even the most unlikely occasions can provide something of interest, and so it was on the *Empire Ken*. First there were the flying fishes. These would rise in silvery clouds, and go skimming like giant marine dragonflies before the advancing bows of the ship. They had a strange, almost locust-like flight and, never having seen them before, I found them fascinating. They must have been capable of rising to some height because a number of the more ambitious ones, airborne to escape what they doubtless thought was the father and mother of sea monsters, managed to find their way to the deck, where they lay wriggling and flapping their enormous pectoral fins. The sea birds of the Mediterranean were also quite new to me. Toward dusk we could sometimes see the dark files of Mediterranean shearwater, weaving bat-like close above the waves; or more rarely the larger Cory's shearwater, sailing and gliding almost like an albatross, then rising to a considerable height to rest on the wind with matchless ease and skill before plunging to the sea to seize some small fish unwise enough to surface momentarily.

In the cool of the evening we left Port Said behind and began the long gentle glide down the Suez Canal to the Red Sea. I was amazed how narrow the canal appeared to be. In places one felt it to be more by luck than navigation that the ship managed to get through at all. This was for me one of the high spots of the voyage. I sat on a hatchway as the sun slipped quietly down the sky, turning the immensity of desert from a golden sea to violet, then to deep rich purple as the shadows came down over the rocky vastness of windswept, sunblasted emptiness. Here and there along the canal banks fellaheen could be seen cultivating narrow strips of fertile land, their camels looming huge in the dusk. The whole atmosphere was one of unimaginable antiquity, unchanged for thousands of years. I thought of my father and of my grandfather, and of how they too had passed this way en route to India many years before, and I wondered what they had thought, and how the scene I had been watching with such feeling had appeared to them. I looked at Tawny, and wondered if he was the first representative of his species ever to travel this way. Kestrels do sometimes spend the winters in tropical Africa, but tawny owls are strictly European and pretty sedentary. Tawny belonged to the British subspecies; this differs from its continental neighbors by the rich reddish-brown color of the plumage, which in the latter is of a much grayer tone. Tawny himself seemed unconcerned with his unique status as Britain's one and only tawny owl ambassador to East Africa, and went on spraying the deck with water from his bath. Cressida looked upon bathing as a duty to be completed as quickly and with as little fuss as possible; whereas Tawny, the hedonist, would wallow like a small feathery hippopotamus until he was virtually unrecognizable. But the speed with which he dried out and became an owl once more had to be seen to be believed, and he kept himself spruce and in perfect condition.

We reached Aden on May 14th, my birthday. Here some of the troops disembarked, along with a number of civilians, including some of the would-be Ground-Nut scheme recruits who, believing that the ship had reached Mombasa, their destination, sobered up just sufficiently to stagger ashore and then, being still absent when she sailed, were seen no more. To me Aden will always mean kites. These big Red Sea kites were everywhere; they came out to meet the ship in hordes and perched gull-like in the rigging, but their agility in the air would have put any gull I have seen to shame. Indeed the black-headed Hemprick's gull, almost equally common, seemed clumsy by comparison. The kites, which appeared completely fearless, would glide about overhead until some morsel was thrown up, when one would perform a magnificent corkscrew-like stoop and clutch the tidbit with its large yellow feet. I thought it would be fun to tame and train a young one just for the pleasure of seeing its dexterity as and when one wished.

Our stay was short and Aden soon lay astern of us. As we skirted the horn of Africa we could just see far off on the starboard bow, shimmering in the heat haze, the arid hills of what had been British Somaliland, but was now part of the Republic of Somalia; they rose starkly savage, but fascinating, just as my father had seen them when he arrived at Berbera over thirty years before at the outset of the campaign against the Mad Mullah, which was to lead him to Government House and finally to death in action against the fanatical dervishes. I had been told, and had read and dreamed so much about all this that I felt almost as if I were coming home, or to start a life for which my previous twenty-seven years had merely been a preliminary training. It seemed appropriate that I should share these moments with Cressida, who had already shared so much of my life and who had kept my hope alive when life was at its bleakest. These thoughts, however, were soon set aside as we put far out to sea, heading for Mombasa and journey's end.

2

The Tanganyikan Farmer

AT LAST the morning came when we awoke and went up on deck to see, almost within walking distance, the low-lying, mangrove-fringed, deep green coast of Kenya, as the *Empire Ken* slowly nosed her way into the harbor of Mombassa.

The heavy luggage was taken straight from the ship to the train which was to carry us on the first leg of our journey to Dar-es-Salaam due south on the Tanganyikan coast. I carried my suitcase and the basket with Tawny and Cressida, but before I was able to leave the quay we were herded into the usual featureless sheds which, the world over, are generally the first and last things the wanderer encounters, no matter how or why he travels. I was somewhat alarmed on being approached by an official-looking figure in immaculate khaki shirt and shorts; he informed me that he was the local government veterinary inspector and had been informed that I had "illegal immigrants" with me. Having little choice, I produced Tawny and Cressida, who perched happily on the Customs' counter, glad to be free and able to stretch their legs and wings. The vet took one look at them, asked me a few questions about their origin and destination, and wrote out an entry permit on the spot. At last my worries about my two small fellow travelers were over. Jock, my ex-cabin mate, was going on to Nairobi to take up an appointment, so we bade each other farewell after promising to meet again when the chance arose.

The first birds I saw were sitting on the roof of the railway station. I thought at first that they were jackdaws, and felt a warm glow in recognition of what I took to be old friends. Then I took a closer look and saw that these birds were bigger, with larger beaks, and less gray on the nape; some time later I discovered they were Indian house crows, descendants of birds that must have been imported and liberated years before, presumably by expatriate Indians or Anglo-Indians, homesick for the noisy, thieving but cheerfully impertinent birds, so much a part of the land they had left behind. Very soon afterward I saw some authentic African birds, a group of large black-and-white crows perched on the dead branch of a tree overhanging a piece of waste ground. They were the size of hooded crows and marked like them except that they were pure white where the hoodies are gray. These pied crows proved to be one of the omnipresent African species, for they are to be found from sea level right up to the farming district of the White Highlands.

At last we installed ourselves on the train that was to take us to Moshi, where we were to spend our first night. This train, from its appearance, might well have been in operation at the time when Colonel Paterson was outwitting the famous man-eaters of Tsavo, the lions that had held up construction of the Mombasa-to-Uganda railway at the end of the last century. Slowly, jerking and lurching, the train steamed its way out of Mombasa, past the banana plantations where the palm-like trees with their deep green fronds and bunches of rich yellow fruit looked even more colorful by contrast with the warm red of the soil. Here and there were the conical, thatched mud-walled huts of the natives. The whole scene appeared strangely familiar, made so no doubt by the films of Cherry Kearton and Martin Johnson that I had watched avidly as a schoolboy just before the war.

I found it hard to realize that I was actually seeing these things for myself at last, and I could not shift my eyes from the window of the train, which was snaking at little more than walking pace through this fantastically rich coastal belt. Soon however the train began to climb and, almost without noticing it, we were moving through a very different landscape. Savage-looking thickets of "wait-a-bit" thorn had replaced the banana trees while the semiaquatic mangroves had given way to the yellow-stemmed acacias. This was the true thorn-bush country —the bush that covers so much of the African continent, and through which the early pioneers and settlers had had to force their way. I looked eagerly for signs of game, but saw none. Nonetheless the landscape had its attractions. Perched high on the very crown of the flat-topped acacias could now and then be seen, silhouetted against the steel-blue sky, large slender gray hawks with long white-tipped tails. These were chanting goshawks, so named for the melodious song which they were supposed to produce during the mating season, though I was never lucky enough to hear them. These handsome, but rather sluggish, hunters of lizards and other small beasts, are typical birds of the dry scrub country, and I was soon to know them well.

Moshi was a small East African township, the few houses owned by Europeans having a curiously Bavarian look about them; this was formerly a German colony. The hotel was run by a hospitable Swiss family, who extended the wealth of their welcome to Cressida and Tawny, and allowed them to share my room and treated them as though such guests were by no means unusual visitors to this somewhat remote part of the world. After settling in and unpacking my suitcase I took Cressida outside and gave her a flight to the lure, which I had

The Tanganyikan Farmer

managed with some difficulty to make for her. She was not particularly interested in the lure, this being her first flight of any consequence since leaving Sussex; she was content to glide and sail about, flickering around and over the strange exotic trees, eventually planing down to land, butterfly-like, on my fist. I was relieved to see the journey had done no harm to her health or temperament.

Tawny, too, though his clipped wings prevented him from flying, scampered about between the tables and chairs, and was at least able to jump and flap from the ground to the back of a chair and down again to the floor. He would let off steam by dancing among the flower beds, and what flower beds they were! Practically every plant that is associated with the tropics was represented; scarlet poinsettias shared the honor with bluish-mauve jacaranda and the deep royal purple of bougainvillea. The hotel walls were camouflaged beneath a mass of convolvulus, and their great trumpets were much frequented by sunbirds. These were incredibly beautiful creatures with multicolored plumage, mainly scarlet, purple, green and iridescent bronze; their long, slender, slightly curved bills were ideally designed to probe deep into the heart of the blossoms to reach the nectar and minute insects on which they live. I watched a pair of these exquisite creatures feeding and they allowed me to approach to within a few feet without showing alarm. The male was gold and green, with a scarlet patch carried like a shield to guard his chest, the female quietly attractive in pale grayish-olive. They repeatedly visited a thick clump of creeper on the wall of a deserted outbuilding. Approaching as close as I dared I heard a faint sibilant, almost cricket-like chittering cry, and on parting the tendrils I saw one of the most perfect nests imaginable, delicately formed of dry leaves and lichen, held together with gossamer, with a tiny opening beneath the roof that projected slightly, giving complete security. In the opening I could see two pairs of jewel-bright eyes and the sickle beaks of the two youngsters which their parents, oblivious of my presence, were cramming full of greenfly.

For my first day in Africa this wasn't a bad start.

Later we sat outside the hotel drinking beer, as the moon rose illuming the great rounded shape of snow-capped Kilimanjaro. The mountain seemed almost within touching distance, crouched like a huge plum pudding, the resemblance being enhanced by the snow which gave the effect of caster sugar in the moonlight.

Next day we climbed into a bus for the two-day journey to Dar-es-Salaam. How well I remember those buses—generally Indian-owned and driven by Africans—that bumped and crashed their way all over East Africa; they were as often as not the only way of getting from point A to point B. The European passengers, if any, occupied the first few seats behind the driver; and the Africans, frequently accompanied by wives, children, hens and even the odd goat, sat behind. The only concession to the *bwana*'s comfort was that the front seats had some pretense at padding, an amenity totally lacking in the others. What always amazed me about these antediluvian machines was not that they arrived late, but that they arrived at all. On this occasion we all piled in excitedly and I, with Cressida and Tawny in their basket on my knees, sat in the front seat just behind and to the left of the driver who, somewhat to my consternation, was barefooted. After a good deal of badinage with his village chums he climbed into the driving seat, the gears let out some excruciating shrieks, and, after a few muffled bangs from the engine, the bus suddenly lurched forward and we were off.

Tanganyika was then one of those parts of Africa which had remained unchanged with the passage of time. On our journey to the Indian Ocean the road, which in many places hardly deserved the title, was skirted by rank tropical forest, and behind this natural screen my imagination saw all manner of strange and savage beasts lurking. We came to a dry gully which, with the onset of the rainy season in a few weeks' time, would become a raging impassable torrent, but now was merely a rift in the uneven rocky surface of the track. On the far side of this fissure was a large troop of big olive baboons which are so much a part of the African scene. These monkeys certainly damage crops and are unpopular with the European settlers. Although their sinister reputation as killers of newborn lambs may be so, I have always found them quaint, humorous and by no means unattractive animals, with a highly developed social system and an unusual willingness among the adults to sacrifice themselves if necessary for the safety of their families.

This group was slowly and methodically crossing the road ahead. The females, some with tiny youngsters riding on their backs, together with a number of the younger males, were in the lead; the rearguard was held by an enormous male of quite terrifying power, with a huge rather dog-like face, beetling brows and small watchful reddish-brown eyes obviously summing up the situation. He opened his mouth to bark out some uncomplimentary remarks, showing teeth far larger and stronger than those of any dog, and then followed his family with a wonderfully insolent rolling swagger. With his ridiculous tail cocked in the air he glanced round once more just to make sure that all was well before plunging into the thicket on the far side of the road, that had already swallowed up the rest of the party. How our ancient and, to all outward appearances, almost moribund bus managed to negotiate these ravines, only St. Christopher knows, but negotiate them she did; the driver, for all his lack of footwear, must have known his job.

A great deal has been written about the birds of Africa but, without actually seeing them in their natural setting, it is impossible to acquire anything but the haziest idea of their incredible variety. Whereas the British ornithologist may, if he is exceptionally lucky, see perhaps once in a lifetime a vagrant example of the roller, here there were at least half a dozen species of these splendid birds, of which the most common and perhaps the most brilliantly plumaged was the lilac-breasted roller. This conspicuous bird, with its fast buoyant flight, something between that of a jackdaw and a falcon—with a touch of tumbler pigeon thrown in for good measure—was a well-blended mixture of warm reddish-brown, green, violet and, as its name implies, a wash of deep clear lilac from throat to breast; and to add further distinction it boasted a clean-cut deeply forked tail. Another group of birds, well represented here, was the kingfishers, varying in size from the big, handsome pied kingfisher of the coastal creeks, to the tiny well-named pygmy kingfisher, a minute insectivorous species, which in common with many African kingfishers prefers dry bush country to the banks of the most enticing streams.

The names of the small towns were full of the promise of adventure and excitement: Arusha, Korogwe and Morogoro. The very mention of these names stirs my blood to this day. The African twilight, brief but intensely moving, was closing in as we approached Dar-es-Salaam, the Arab's haven of peace. A few miles inland, silhouetted against the darkening sky, I saw a solitary hunched figure perched among the top branches of a huge baobab tree; from its head, as it turned to stare at us as we clattered past, protruded a pair of semierect horn-like ear tufts, which gave their owner a sinister almost Mephistopholean appearance. It was an eagle owl, probably from its size the great Verreaux or milky eagle owl, the largest and most powerful of all the East African owls.

The few days I spent in Dar were busy ones. Luckily I had made friends on the journey with a young couple who lived in a lovely old house just outside the town, and here I passed two evenings before continuing my journey to Iringa, more than two hundred miles to the southwest. While we were sitting on their verandah drinking a sundowner, the humid, breathless night air was suddenly shattered by a shriek, such as might have been uttered by a tortured child. This was followed by another and yet another. My host and hostess smiled and looked quite unperturbed. "Oh, don't worry," they said. "It's only the bush babies. You'll soon get used to them." There was a mango tree in full fruit at the bottom of the steps leading to the verandah that was lit from the porch close by. Suddenly a pair of slender leaping shapes appeared, ghost-gray in the shadow-dappled fruit tree. Daphne, my hostess, quietly and quickly reached for a powerful spotlamp and switched it on for a few seconds. The beam caught and held one of the most enchanting creatures in the whole of Africa; the size of a half-grown kitten, with squirrel tail, tiny hands, a pointed almost cat-like face, and huge eyes set off by great bat ears that furled and unfurled to catch the slightest sound. All this I saw before the bush baby, still gripping a ripe mango in its mouth, leaped for an overhanging branch and, together with its mate, vanished so swiftly and completely that I half wondered if it had ever been there at all.

After an almost interminable journey by train and bus we arrived at Iringa. The club at Sao Hill in the southern highlands of Tanganyika, about sixty miles from Iringa, was the center of the European farming community, being roughly the Tanganyikan equivalent of Kenya's White Highlands. Here Lord Chesham, following the example of his fellow peer and visionary, Lord Delamere of Kenya, hoped to found a colony of expatriate British settlers, to bring British methods of agriculture, animal husbandry and, more especially the British way of life to this unexploited equatorial paradise.

My original purpose in coming all the way to Tanganyika was first to learn how to manage a farm, and then, if I enjoyed the life, eventually to buy one of my own. It did not, I fear, take me long to realize that I was not really keen on farming. As things turned out this was just as well because I believe that, after the Mandated Territory of Tanganyika changed its name to Tanzania, a good many settlers were dispossessed of their land and had to return to Britain, leaving all they had dreamed of and worked for behind.

When I eventually arrived, tired, travel-stained and thirsty, my preconceived idea of the white man in the tropics was soon confirmed. I fear that I must have set the older, more established and, above all, more conventional settlers back on their heels. To start, when I first entered the club carrying Cressida on my fist and Tawny in his basket, I was regarded with the gravest suspicion. This Just Wasn't Done. Most of the earlier arrivals were already farming their own land, and had formed a little community of their own. The majority had been field officers at least, and had always led somewhat ordered lives; they just didn't know what to make of me. Besides, to make matters worse, I showed a lamentable lack of enthusiasm for cricket.

After a few days, during which it was obvious that my movements were being watched and reported upon, I was sent as a so-called pupil to a local farm, where I was supposed to learn the rudiments of tropical agriculture. For this privilege I was asked to pay no less than twelve pounds per week, and I found to my horror and amazement that I was expected to take over the virtual running of the establishment. My duties consisted of supervising the African labor force, seeing to the welfare of the stock, maintaining the fire breaks and doing a great many other things besides. At the end of my first week it became increasingly plain that my employer knew even less about farming than I did. There was, however, a subtle difference in our positions, for he had invested thousands of pounds in his rolling, well-watered acres of farmland, whereas I had contributed nothing; moreover he was very seldom in resi-

The Tanganyikan Farmer

dence. He preferred the company of his fellow exiles, who could generally be found either at the club itself or attending settlers' meetings in Iringa, which it would appear were highly convivial occasions. Anyhow he seldom put in an appearance, and I had the place almost entirely to myself, a state of affairs that suited me well.

The farm was a huge affair of several thousand acres, divided into immense fields of which the main crops at that time were maize and sunflower. All the haulage and plowing was done by oxen, gentle hump-backed beasts, obviously the result of crossing imported Indian zebus with some type of native cattle. They had a formidable sweep of impressive horn but they were utterly lacking in aggression and I soon became fond of them and knew most by name.

The country was vaguely reminiscent of Wiltshire or the Berkshire Downs, at an altitude of about eight thousand feet; it was well irrigated, and the day temperature was that of an English summer, though it dropped to a little above the freezing point at night. The labor was recruited from the local tribe, the Wa he he. Now friendly and even humorous fellows, given to over-indulgence in *pombe* (the native home-brewed beer) and dancing, they had at one time been warriors of some note. When under German administration, before the First World War, they had mounted a short and ill-fated attempt to throw off alien domination. Apparently they had consulted their medicine men who had told them that if they uttered certain magic incantations the bullets of their oppressors would be turned into water. Believing this they rushed fearlessly upon the foe, only to be mown down by the hundreds. This abortive attempt to obtain independence has been called the *Maji-Maji* (Water-Water) rebellion ever since.

The life itself was pleasant enough. I would get up in the pearly African dawn, just as the sun was clearing the belt of wattle trees that sheltered my little reed-thatched "manager's" house. My houseboy would bring me a strong cup of tea and the native groom or *sais* would saddle one of the several ponies; these were my only means of getting about, as much of the farmland was hopeless for any sort of wheeled transport. My favorite pony was an Australian whaler, appropriately named Gundergai; he was a bright bay with the endearing habit of remaining nearby while I worked at a job that demanded my full attention. I would trot and canter about the estate, supposedly supervising the African farm workers and making sure that they were busy about their tasks, but my eyes and ears were always on the alert for more exciting subjects. Perhaps I would catch a glimpse of the furtive gray shape of a jackal returning to his den in an aardvark's deserted hole; or I might disturb a small group of reedbuck who, whistling shrilly with alarm, would go bounding stiff-legged to the cover of the nearest ravine.

I was relieved to find that, on the whole, the farm ran itself satisfactorily with a minimum of interference from me. The Africans were under the direct control of a native foreman or head man, called the *Neo-para;* he had served as a sergeant-major in the King's African Rifles and knew a great deal more about running the place than I did. I would return at about nine o'clock for a breakfast of eggs and bacon, together with many cups of tea. Afterwards I would take Cressida for a tour of the farm and give her as much flying exercise as she wanted.

There were all sorts of other small falcons and falcon-like hawks, such as lesser kestrels, red-footed falcons, hobbies, both African and European and, perhaps most attractive of all, the gull-like black-shouldered kites, small white birds no larger than kestrels, with deep inky patches on their shoulders. A pair of these delightful birds was nesting in an isolated group of acacias in a remote corner of the farm, and they could be seen every day in the late afternoon scouring the open savanna for their prey, mainly locust-sized grasshoppers and field mice.

These mice were pretty creatures, neatly striped, and looking exactly as though they were wearing little black-and-white school blazers. The adult kites would hover rather like kestrels but, when about to drop on to their prey, would raise their wings full stretch above their backs and lower their legs like an airplane's undercarriage before parachuting quite slowly into the thick grass. This method must have been effective, as I seldom saw them rise empty-footed. If I cast off Cressida anywhere near their home they would come dashing out, stooping at her hard and often, while uttering a peculiar, almost hissing, whistle of indignation. Cressida, outnumbered and obviously embarrassed by this unseemly show of bad manners, would veer off and sail away downwind until near her own territory, when she would sometimes retaliate and put in a few stoops of her own, just to show that she refused to be bullied too much.

Meanwhile Tawny, who had begun to moult and with the growth of new feathers had developed considerable self-confidence, was beginning to find his wings at last, and as the blue-gray owl light stole silently over the resting farmland he would descend from his perch on the mantelpiece where he spent much of the daylight hours, his great dark eyes would appear even bigger in the dusk, and he would half-fly, half-dive through the open window of my sitting room and disappear for a bit of solitary exploration. He never went far, and would reappear at different windows, peering in and rotating his head in a curious manner, making sure I was still there, perhaps to remind me that his supper was due. He also discovered his voice, and it was a strange nostalgic experience to hear the long quavery full-throated hoot of a tawny owl coming from out of the gentle moon-silvered African night.

One thing was lacking and strangely enough it took me some weeks to realize what it was; then suddenly it dawned on me as I lay in bed listening to the far-off whooping hunting calls of a spotted hyena. The voice of the hyena is to me the voice of Africa, haunting, melancholy and wild. What the call of the curlew is to the wild places of Britain so is the voice of the hyena to the African bushland. I heard it again, getting closer, full of desolation and an intense loneliness. I felt lonely too, but I knew what I needed—the company of a dog. The farm workers employed on the estate lived in what amounted to a small self-supporting village with their wives and children in the usual mud-walled huts common to the whole continent; here they kept poultry and grew a few crops; they also kept dogs.

These dogs belonged to a variety known unflatteringly as *pi* dogs or pariahs. Although certainly not a breed in the accepted sense, nonetheless they have certain pronounced characteristics. They are usually a warm reddish-yellow, with pricked ears and bushy tails. Their prototype is the dingo of Australia, itself believed to have been an immigrant from Asia, arriving with the ancestors of the aborigines many centuries ago. The dogs at Muheka farm were not the half-starved pathetic curs one sees so commonly hanging about the bazaars of the Middle East. The Wa he he were a tribe imbued with strong sporting instincts and above all they took pride in hunting with packs of dogs. Consequently the dogs were normally well cared for.

I knew that one of the grooms had a litter of puppies, sired by a dog named Kali (the fierce one), boss dog of the Muheka farm. I duly presented myself outside the hut where the groom lived and asked to see the puppies. He led me to a sort of lean-to, cleverly made out of interwoven dried maize stalks, to

which the great broad leaves were still attached, the whole contraption making a most satisfactory shelter. As he whistled a bitch emerged looking both nervous and embarrassed; native-owned dogs seemed to have a natural suspicion of white men. She was followed by four fat little puppies; three looked as bashful and shamefaced as their mother; the fourth, the biggest and fattest of the litter, sat down on his little round behind, cocked his head on one side and leered at me insolently. He waddled forward, looked up at me, and gave one loud clear *woof*. I picked him up and put him, fleas and all, inside my shirt where he nestled close to my body, handed over ten shillings, and mounting my pony returned as fast as possible to my house.

The puppy, on close examination, looked like an overgrown bumblebee; his soft baby coat was a deep warm reddish-gold, save for his spotless snow-white shirt front and a curious dark saddle-shaped patch extending from his withers to his rump, which later gave him a most distinguished appearance. His bushy tail curled over to meet his hindquarters exactly like the handle of a jug, and he had deep creases in his forehead that gave him an air of wisdom and thoughtfulness, rather pathetic in one so young. He was just six weeks old. For no particular reason I called him Rupert, and the name seemed to suit him perfectly. It was soon obvious that he supported a large and varied collection of insect life and needed a good dusting with pyrethrum powder. This was easily arranged; we grew acres of the stuff, and one of my first impressions on arrival at the farm was of being dazzled almost to the point of blindness by the sunshine reflected from field after field of these brilliant white daisy-like flowers. Incidentally the natives wore sun glasses while cultivating this crop.

Now I had a dog of my own once more, and life for me was complete.

3

A Pup Named Rupert

I HAVE ALWAYS LIKED the unusual and Rupert was certainly no ordinary dog. His ears, unlike those of most of his tribe, were not fully pricked but turned over at the tips giving him a gentle, kindly look which his nature did not belie. From his earliest days with me I used to carry him around the farm tucked into my bush shirt, and he must have traveled many miles in this way. As he became stronger he would follow my pony, and could be seen plowing through the tough wiry grass, which was a great deal taller than he was, like a small sturdy fishing boat battling with outsize waves. We spent a great deal of time together, and he soon proved his worth as guard, companion and hunting partner. Although he came from a much-despised race he was loyalty and courage all through.

The Africans, in common with the Welsh, have a gift for applying appropriate nicknames commemorating some outstanding characteristic or idiosyncrasy. Because of my interest in birds, I quickly became known as *Bwana N'dege* or Mr. Bird Man. Furthermore, they soon realized that I was a soft touch when it came to selling animals they happened to come across in the course of their wanderings about the bush. The first to arrive was Swara. I was reading in my sitting room during the heat of the day when I heard the voices of a group of Africans. Going outside I saw that the leader of the group was carrying a bulging, colorful calico bundle. I had learned a fair amount of Swahili by this time, so I asked him in that language what it was. "*Nyama, bwana*" (an animal, sir), he answered, and coming closer I saw a tiny pointed face with great liquid dark eyes gazing pleadingly up at me. It was an Oribi fawn, about ten days old at the most. "What do you call this?" I asked. "*Swara, bwana,*" he replied. Swara is the Swahili name for a number of small buck. The Africans, being hunters rather than naturalists, were more concerned with an animal's edibility than its species. I pulled out a ten-shilling note and the buck was mine. I also made it plain what I thought of them for trading in small animals in this unscrupulous way, but I doubt if I made much impression.

Swara was a wholly delightful creature, exquisitely formed, with long slender legs ending in minute black hooves, the size and shape of a fingernail. Her body, as light and slender as a whippet, was clad in reddish-gray hair with the texture of plush; her neck was long and gracefully arched and indeed her whole appearance was almost enough to turn me into a vegetarian on the spot. Placed on the hard slippery stone floor she nearly did the splits, so I took her to my bedroom which was covered with hides, a legacy from the previous occupant. Catching sight of Rupert she bleated shrilly and trotted forward to investigate; being hungry, and Rupert being the nearest in appearance to the mother that she had not seen for some time, she searched him hungrily for sustenance. Rupert, then about three months old, was puzzled but friendly and willing to cooperate; too much, however, was being expected of him. Instead I managed to get hold of a baby's bottle and a plentiful supply of goat's milk; and a visit to Iringa later produced a bottle of lime water which was added to the milk.

A Pup Named Rupert

Swara throve and seemed to grow almost hourly. She was tame from the start and soon became both playful and affectionate. She would trot and canter after me on her tiny little legs, and would try to engage Rupert in butting matches using her bony little forehead with great enthusiasm; he, for his part, though obviously embarrassed by these attentions, never lost his temper with her. She loved nothing better than to lick my hand with her rough tongue, probably in an attempt to satisfy her longing for salt which is an essential part of the diet of young antelopes. Later I overcame this lack by filching pieces of rock salt destined for the farm stock. She would sleep curled up at the foot of my bed, awaking at intervals during the night and chew my ears and nuzzle my neck to remind me that she was hungry. Fortunately she was soon weaned, after which she acquired a passion for the leaves of tomato plants, and would come careering into the house with several feet of this succulent greenery trailing from her champing jaws.

She was quite popular with my own personal staff, but I had to be constantly alert lest some wandering stranger with a taste for venison might carry her off. Later she became increasingly nocturnal, disappearing into the bush at dusk, to reappear in the small hours, tripping into the house, her hooves sounding like the steps of a tiny tap dancer. She would join me at breakfast, where she showed a most unnatural predilection for fried bread and bacon rind, and would then sleep away the sunlit hours in the coolest spot she could find. As she grew older she would disappear for longer and longer periods, and I have every reason to believe that she eventually returned happily and successfully to the bush from which she had been kidnapped. I was certainly the happier for my brief contact with this charming animal.

Some time later I rescued a dik-dik that had been netted young and was on the way to enrich some local Wa he he stew. This was an even smaller antelope than Swara and comical looking, having a rudimentary proboscis like that of a tapir and a pronounced almost Hitlerian forelock. Unfortunately it was not possible to tame him properly as he was already fully weaned, so I took him to a part of the farm where I knew a family of dik-dik lived and sent him on his way.

One of my duties was to supervise the dipping of cattle; this had to be carried out regularly so as to keep them reasonably free from disease-spreading ticks. Although I realized this job was essential, I disliked it intensely, finding it hard to remain indifferent to the terrified bellowing and desperate milling about of the frightened beasts, which had to be forced down a ramp by prodding with long poles until they finally lost their footing, plunged into and swam the length of the bath. However it was obviously effective as few of our cattle succumbed to any of the prevalent fevers.

On my way to the dip, which was far away on the very perimeter of the farm, I had to pass a large isolated thorn tree of considerable height and as flat-topped as a table. High in the branches was an enormous nest, obviously that of some bird of prey. I happened to mention the whereabouts of this nest to one of the night watchmen, whose job it was, among other duties, to look out for and report any signs of fire, either on or near the estate. Seeing my interest he offered to climb the tree, and up he went with remarkable speed and dexterity. When he was about halfway up there was the devil of a commotion overhead and suddenly an immense lappet-faced vulture, probably the largest of all African birds of prey, rose from the upper branches, unfurling its nine-foot spread of sail to circle overhead in great lazy spirals—a superb sight. The nest, on which I could have stretched out comfortably, contained one dirty white egg nearly as long as my hand. I told the night watchman to replace the egg and, after retiring to some distance, had the thrill of watching the colossal bird drop out of the sky and shuffle back on to its nest to resume brooding.

A Pup Named Rupert

Two or three months later both parents were dead, accidental victims of a poisoned bait put out by a neighboring farmer for the destruction of a marauding hyena. Once more the watchman (whose name, incidentally, was *Kasi N'Dogo* or Little Work, a title that suited him admirably) climbed to the nest, which contained a young vulture some weeks old but still completely covered in grayish-white down, except the head which was hideously devoid of either fluff or feathers. It must have been without food for several days and was very weak; nonetheless it was all I could do to carry it the mile or more to my house. The bird, on close examination, had features of fascinating ugliness, and for some strange reason brought to mind Mr. Bultitude, hero of the Victorian school story *Vice Versa*. As Bultitude seemed to me rather an awkward name, I had to think of an alternative; after several tries I hit on Maloney, and Maloney he remained as long as we were together.

He was the easiest bird in the world to feed. All I had to do was to prize open his heavy horn-colored beak, ram great lumps of meat down his throat and massage his skinny neck; he would then give a couple of gulps and that was that. His head and beak, which always seemed much too heavy for his long goose-like neck, would rest comfortably on a sort of pillow I had made for him, and having eaten his fill his face would assume the supercilious expression of a camel. He was not exactly attractive, but I became very fond of him and it was fun to see him shuffling forward on his stomach to receive his rations, his legs being much too short to support his huge body. When hungry, which was almost always, he would give vent to a ridiculous high-pitched whistling squeal, and when he saw that nothing further was forthcoming he would grunt like a surly hog. He spent a lot of time asleep, and when awake he would turn his long, almost naked neck and preen his embryonic wings and the ludicrous tuft of down that would one

day become a tail. Fortunately meat was both cheap and plentiful, otherwise the cost of supporting such a glutton would have been prohibitive. As it was he grew in bulk and strength, if not in beauty, at an alarming speed and could soon stand up and walk slowly about the verandah, every now and then falling forward on his breast, when he would toboggan along using his wings as well as his feet to help him forward.

Maloney was not popular with my visitors, either European or African. One evening a hard-drinking Afrikaans ex-gold miner, who had never seen or heard of him, called in to help me finish a bottle of brandy and spent the night, as was the custom in that part of the world. He had gone outside shortly before turning in and was suddenly confronted with this apparition perched upon the back of a chair on the verandah. Almost inarticulate with horror he staggered inside and swore to lead a better life in the future, a resolve that I much doubt he kept for more than a brief period.

On another occasion, when Maloney had learned to become airborne, I decided it was time he made his debut in society. With considerable ingenuity I made a hood of the type used in falconry, but with one important difference; a large hole was left open on either side of the hood to allow Maloney's large clear blue eyes to peer unobstructed upon the world. I then covered the hood with black cockerel feathers and, after a good deal of gentle persuasion, induced Maloney to wear it. Thus disguised he really looked very impressive and it was hard to believe that he was not a magnificent eagle with an outsized beak and almost frightening stretch of wing. After a struggle I managed to hoist him onto my left fist, where he squatted happily enough, his shoulders about on a level with the bridge of my nose. One of the ponies, appropriately if unromantically called Block, was as powerful and phlegmatic as a hippopotamus and showed little concern when I struggled somehow into the saddle. Block broke into his usual ambling trot and, despite a bit of balancing trouble during which Maloney completely blotted out my view with his nine-foot wing span, we soon arrived safely at the Club.

Our reception was all I could have wished for. It was Saturday evening and everybody who was anybody was there. They all crowded round to admire this magnificent raptor who sat immobile as a rock upon my aching arm. Someone ordered me a pint of beer and I settled Maloney with some panache on the back of an unoccupied chair. I took a long draught and turned to Maloney just in time to see him insert his hind claw into the space between his hood and neck, and to my horror, with a quick flip, jerk it off, seize it in his beak, and toss it into the air. He then caught and swallowed it, feathers, leather and all; he cocked his head on one side, deliberately closed one eye, and gave me a slow lecherous wink before stretching out his head and neck, naked and obscene as ever, to pinch a sausage from the plate of one of the members. As will be gathered Maloney had a well-developed but decidedly perverse sense of humor.

There were a number of rondavels used for putting up casual guests. These were round mud-walled huts, with pointed thatched roofs somewhat like miniature oasthouses in shape. Maloney was adept at discovering which of these rondavels was occupied and at first light would come winging in from his roosting place on the top of an outhouse. He would then alight on the top of the hut, looking like an enormous animated weathervane until, losing his balance, he would slide flapping and scrabbling down to the ground. The occupant, aroused by the hubbub, would rush outside to be confronted by Maloney who would put on a terrific display of bogus ferocity. He would flap his huge wings and waltz around in a sort of war dance, accompanied all the while by a series of loud yelping barks. This was of course pure bluff for he was never known to hurt anybody, but it did little to boost his position in the popularity stakes.

Shortly after, I met a very remarkable man. When I met Captain Livingstone Stilling, the owner of the next farm but one, he must have been well over seventy, but he still ran his huge farm virtually single-handed. He rode a strong-willed black Arab stallion called Ensio, and could spend hours in the saddle, attend to his accounts in the evening, and then drive his dilapidated old Chevrolet to the Club for a night's carousal. An Australian by birth, he had been a breaker of brumbies, the Australian equivalent of the American mustang, and had sailed as a member of the crew of a tea clipper. He had fought in Russia in the First World War and been commandant of a Polish refugee camp in the Second. In addition he was an extremely competent farmer and ran one of the best land units in the district. Although a long-term diabetic he was very fond of his liquor. On one occasion shortly before his death he was warned by his doctor to cut out the drink. According to legend he poured himself a stiff brandy and replied, "Well, doctor, it's better to die of D.T.'s than diabetes!" He was widely known as Stilly, the Churchill of the Southern Highlands, and indeed he bore a remarkable resemblance both physically and in mannerisms to that great man. Not long after we met, Stilly offered me a job as assistant manager on his farm; he also transferred from his other farm his manageress, an extremely competent, hard-working but nonetheless amusing and highly sociable German lady, who undoubtedly knew as much about tropical farming as any man in the area. Ula had spent many years in Tanganyika and spoke both Ki Swahili and Ki he he like a native.

Having accepted the job I knew that before long I would

be involved in a confrontation with the redoubtable Ensio. It came almost at once. The very first morning, when I presented myself for duty, Stilly asked me to ride over and have a look at a stretch of irrigation furrow that had caved in under the trampling hooves of a small sounder of bush pigs. He added that, as it was a long way from the house, I had better ride Ensio, who was already saddled and hitched to a rail nearby. Hardly wanting to refuse I approached this fearsome beast who was standing quietly enough resting one hind leg, with his splendid tail idly swishing at the ever-present horseflies. Ensio was blacker than night, with the aristocratic face and small muzzle so characteristic of the Arab breed. His eyes were large and intelligent and his powerful arched neck, glossy as a rook's wing, bore evidence of his unimpaired masculinity. As I took the reins from the native groom and gave Ensio what I hoped he would take to be a confident pat on the neck he turned his head and gave me a long speculative look, summing me up no doubt. Watched by Stilly and a small group of Africans I swung nonchalantly into the saddle, collected the reins, gave a slight squeeze with my heels and awaited the expected explosion. Ensio cocked his ears, arched his neck, gave a gentle whickering sound and strode quickly and smoothly out of the yard.

This was fine as far as it went, so taking my life in my hands I gave a slight kick with my heels, at the same time making what I hoped were friendly and encouraging noises. Ensio lengthened his stride and broke into a long, smooth desert-devouring trot. This was terrific! Ensio must have recognized a natural horseman when he saw one, a quality so far unsuspected, especially by myself! Soon we were out on the open savannah, and my confidence began to grow. Surely this gentle, almost lazy little horse could not be the fire-eater I had heard so much about. I gave him another tap with my heels and he responded at once; the trot gave place to a canter and we were floating over the veldt grass, dodging the antbear holes, the morning heat disappearing at the speed of our going. A Kori bustard, bigger than a turkey and with great powerful wings, lumbered out of the grass ahead of us and made off gliding low over the ground. I relaxed as I watched its slow flapping flight and Ensio saw his chance.

Breaking into full gallop he shot ahead with all the speed he could pile on; after traveling for about fifty yards he stopped dead in his tracks, put his head between his forelegs and shook it, boring like a hooked salmon. There was no answer to that one as I shot over his withers, landing with the airy grace of a winged pheasant right in the center of the only patch of wild sizal for miles. Sizal cuts like razor blades, and I was lucky to escape with only a few scratches on my arms and a few sizable rents in my shirt. As I lay prostrate Ensio gave me a look in which amazement and scorn were equally blended, and then trotted off to graze on a patch of burnt ground through which fresh green grass was just beginning to sprout. Finding that I was more or less intact I gathered myself together, muttering hard things about Ensio and his ancestors for many generations past. I caught him without effort; indeed he seemed quite satisfied with this demonstration of guile and allowed me to remount and ride on without further unpleasantness. At least I was thankful that no prying eyes had witnessed my overthrow. He tried the same trick again a few days later, but this time I was ready for him, grabbed the pommel of the saddle and somehow managed to weather the storm. He had one other trick in his repertoire: finding he could not dislodge me by the old method he swerved suddenly to the left when going flat out. I nearly repeated my first ignominious descent, but I kept my feet in the irons, clutched him round the neck in an anything-but-loving embrace, struggled back into the saddle, and finally brought him to a standstill.

One day Rupert and some of the farm dogs put up a bush

buck ram and coursed it with a burst of music almost worthy of Pytchley. Ensio pricked up his ears and broke into his tireless gliding gallop. I urged him and the pack to further effort, and the chase was long, exciting but fruitless. The bush buck, a wily old fellow, victor of many such hunts, eventually vanished into an almost impenetrable reed bed from which it would have been impossible to flush him. The dogs returned exhausted to flop out in the shade; and Ensio, his sides heaving and his flanks lathered, walked quietly behind me all the way back to the stable. From that day on there was a difference in our relationship; we had discovered something in common, the thrill of the chase, and there was no more trouble. Later I could remove his saddle and bridge and just leave him to graze while I supervised some particular job. He would even share my packet of sandwiches and cake while I took my midday siesta under the shady branches of some convenient acacia, and I knew that he would be at hand to carry me homeward when the work was done.

I had always believed that Africa would be a lepidopterist's paradise, but in this I was at first disappointed. There did not appear to be anything like the variety of species one would find, for instance, in Ceylon or in parts of South America. Nonetheless butterflies were there and in particular those aristocrats of the insect world, the swallowtails and hawk moths, were well represented. One of the commonest butterflies was the tail-less swallowtail which bore a close resemblance to the British swallowtail, except that the hind wings totally lacked the distinctive projections for which the family is named. These large and colorful black and cream insects were everywhere on the open veldt. They were strong flyers but seemed to lack the airy buoyancy of the more orthodox members of their family. There were, however, some very attractive true swallowtails, whose deep velvety-black wings were bisected by narrow vertical bars of emerald green; and these, when seen in the full blaze of the sun as they drank from the shallow water of some partially dried-up stream, never failed to excite me, no matter how often I encountered them in my forest wanderings. These swallowtails, together with various species of clouded yellows, whites, and mother-of-pearl butterflies, used to congregate in flocks on mud banks where they appeared to be sucking up the thick soupy liquid with much gusto. Clear water they seemed to avoid; perhaps they needed some sort of mineral supplement to their usual diet of nectar.

Another group of butterflies that I came to know well were the big dashing charaxes, with their curious leaf-like wings and strange passion for the juices of carrion and fresh animal droppings, a habit shared by the glorious purple emperor of Europe. A very common butterfly on this particular farm, albeit one that I have not encountered elsewhere in Africa, was one that I have never managed to identify. I called it the eavesdropper because of its habit of roosting at night and spending the heat of the day beneath the eaves of stables, cowsheds or other outbuildings. If disturbed, the eavesdroppers would fall in purple clouds, not returning to their dark and cobweb-festooned refuge until one had withdrawn from the area. These were handsome mauve insects, their wings thickly dusted with white and red dots, and about the size and build of a large tortoiseshell although their style of flight, slow and lethargic, was entirely different. Unfortunately I did not possess a book dealing with the lepidoptera of that district. Subsequent visits to the Coryndon Museum in Nairobi did little to increase my knowledge, and it was not until I returned to England that I was able to identify most of the species in which I was interested.

The hawk moths, which visited the beds of sweet-scented

tropical flowers at dusk, were in many cases identical or closely allied to species that had previously thrilled me in Europe. Thus the death's head and the oleander hawk moths were by no means the rarities that they were in Britain; and in addition to these comparative giants there were a number of smaller swift-darting species similar to the various striped hawk moths so sought after by the European lepidopterist. When evening came large numbers of moths were on the wing, together with innumerable flying beetles and the fat, aptly-named "sausage flies," in reality the winged form of the hard-biting Safari ant. I would then light my pressure lamp, and with Rupert at my feet would await the arrival of these visitors from the flower-scented darkness without.

Tawny awaited them too, just as eagerly but with a different motive. As they blundered through the open door or window and crashed against the whitewashed wall, Tawny would drop from the back of his favorite chair and shuffle about like a crab, grabbing with his powerful talons anything that took his fancy. He would hold a beetle in one foot, rip off the head, seize the fat juicy body in his beak and close his eyes in a look of sublime contentment. His aldermanic greed took a lot of satisfying.

One of the most spectacular visitors was the magnificent peacock moth with a wing span only a little less than that of a pipistrelle bat. The purple-brown wings were marked with "eyes" similar to those of the European emperor moth, and indeed the two species are closely related. This moth reminded me vividly of the toy airplanes powered by twisted elastic that I had played with as a boy, which also carried colored circles on their wings. Sometimes, when I was moth-watching on the grass outside the house, I would be joined by another owl, one so minute as to be almost unbelievable, smaller even than the tiny Scops owl. This was the pearl-spotted owlet, whose dash and fearlessness more than made up for its diminutive size.

This little bird would come hurtling out of the darkness into the circle of light cast by my pressure lamp. As I sat motionless it would seize its prey, perhaps a praying mantis or a beetle, within a few feet of me. Later I was overjoyed to find that a pair of these delightful owls were living in a thick shady tree only fifty yards from the house.

On Saturday evenings we would all meet together at the Club. Some would arrive on sturdy, country-bred ponies; others came jolting and rattling up in an amazing variety of ancient and decrepit motor vehicles, many of which were homemade and looked as if they might disintegrate on contact with the first pothole. The farmers themselves were dressed in everything from Savile Row suits to garments that would not have disgraced a Western film. Many of the families were accompanied by their dogs; there were immense muscular ridgebacks, vociferous and truculent terriers, and honest-to-goodness mongrels and pi dogs like Rupert. The dogs lost little time in seeking out their friends for a romp or their enemies for a battle, so the scene was uncommonly lively and the din prodigious.

I always took Rupert along on these occasions because I liked his company on the long homeward ride through the night-shrouded bush, where the silence was broken only by the liquid far-carrying call of the fiery-necked nightjar and the distant whooping of spotted hyenas. Later in the evening a number of us would often drive over to the house of one or another of the families to continue the jollity far into the night. On these occasions I would ask someone at the Club to keep an eye on Rupert until I came back to collect him. As most of the settlers kept a number of dogs as guards or for hunting, I did not normally take Rupert with me in case a fight started in which, of course, he would have been completely overwhelmed. Rupert, however, had other ideas. A number of these farms were two or three miles from the Club building and yet several times, just as the party was in full swing, Rupert would appear jumping through the open window. He would single me out of the crowd and hurl himself straight into my arms, quite unconcerned as to what happened to the contents of the glass I might be holding. This happened too frequently to be pure chance and, as I used to leave Ensio in the Club stable, he could hardly have followed my scent. I can only assume it was some sort of second sight. Whatever the explanation it filled me with a great warmth toward him and he soon became a well-known and popular local character.

One day I received a message that the Polish refugee camp near Iringa was closing down and its inhabitants returning to Europe. It appeared that they had a young eagle as a pet and were anxious to find a home for it. Within a few hours I had cajoled someone into giving me a lift, presenting myself at the camp as a would-be eagle foster-father. The Camp Commandant met us at the gate and showed us the eagle sitting appropriately enough on the roof of the cookhouse. It was a very young Bateleur, just on the point of learning to fly. However, a piece of meat brought him scrambling and flapping to the ground where I picked him up, getting well bitten in the process. I was soon to learn that Bateleurs, unlike the more orthodox *true* eagles, are inveterate biters and do so at the slightest excuse. His name was unpronounceably Polish, so I called him Torquil.

Now I had four birds of prey, Cressida, Tawny, Maloney and Torquil, and I found that this number of carnivorous birds was more than I could cope with. With considerable regret I gave Maloney to one of his admirers, a rich vet who lived some way off and who had both the land and the resources to look after him properly. So one afternoon Maloney was driven away to Mufindi, perched proudly on the passenger seat of an elderly Chevrolet, quite content to change his allegiance. Later I visited him in his new home and he seemed happy with the freedom of the entire estate and the surrounding countryside. His new partner (owner is hardly the right word) was a keen naturalist with a penchant for bizarre beasts, shared fortunately by his wife, and as few members of the animal kingdom were more bizarre than Maloney he was a most welcome addition to the household.

Meanwhile Tawny had completed his moult and with his brand new tail and flight feathers had become a very handsome bird indeed. He used to accompany me on long evening walks, perched hawk-like upon my fist, every now and then sailing away with his soft soundless flight to intercept some insect or small rodent that was unwise enough to emerge from cover. He was becoming increasingly independent and would leave his roost at dusk and go winging off into the gathering gloom. Where he went and what he did was his own business, but he was always back by first light, sitting on his favorite corner of the mantelpiece, drawn up to his full height and peering at me quizzically through half-shut eyes. He was seldom hungry after these nocturnal forays and the fat pellets he produced bore witness to his skill as a hunter. Sometimes at night his melodious quavering hoot would come wafting in through the open windows from some far-distant patch of moonlit scrubland, seeming somehow to complement rather than to compete with the native chorus of cicadas and nightjars and the faint croaking of frogs in the marshes away to the south of the farm.

One morning I found his perch on the mantelpiece vacant and feared that some mishap had overtaken him. Perhaps, with his aggressive courage and total lack of respect for anything in fur or feathers, he had tackled one of the big eagle owls that lived in the district. I rode and walked for miles calling his name and whistling, but to no avail. However next morning he was back on his perch, looking smug and entirely unrepentant. From then on his absences became more frequent and longer until he would be away a week or fortnight. Finally he left home forever. I knew that by now he was self-supporting and a match for most of the smaller predators that he was likely to encounter in the wild. Even then his challenging hoot and loud startling KEE-WHICK could at times be heard as he floated far and wide across the tree-dotted veldt. Tawny owls have lived for thirty years in captivity, so it is just possible that Tawny is still alive somewhere in the African bush, thousands of miles from his homeland. Whatever his fate his life must have been far happier than if he had remained a flightless captive in that backyard in Petersfield.

4

Shark Hunting on Lamu

ABOUT THIS TIME I decided that farming, with the obligation to remain in one spot, was not for me, and I began to look around for another job. Every week a Bristol freighter aircraft used to arrive at the Club airfield—a narrow and none too smooth strip of dried grass and flattened termite heaps cut out of the surrounding bush—to pick up the vegetables grown by the local settlers and fly the load down to Natchengwere, the center of the Ground-Nut scheme. This provided an insatiable market for the farmers' produce and was in fact their chief source of income. There were several of these freighters, owned by two Nairobi-based firms, Airwork and Skyways. I soon knew the crews well as I used to help sort out the cargo and load the airplane, after which we would all converge on the Club for hard-earned refreshment.

One day a relief crew arrived, captained by an immensely tall pilot, known hardly surprisingly as Lofty. Lofty owned a small independent airline consisting of a number of Ansons and other small planes also operating from Nairobi. His particular gimmick was flying up-country residents from the White Highlands of Kenya to the tiny island of Lamu, just off the coast, seventy miles north of Malindi. The passengers would spend the weekend at the only hotel on the island, returning late on the Sunday afternoon.

Lofty had what then seemed a brilliant idea. He wanted to start a shark and crawfish business on the island, and he needed someone to organize it. The idea, as it was explained to me over ice-cold tankards of Tusker ale in the club bar, seemed both simple and exciting. It appeared that the seas on that part of the coast were stiff with sharks, both man-eaters and the smaller but still impressive species that prey only on other fish. At that time shark liver was fetching as much as seven hundred pounds per ton; the dried flesh was in demand as food for African labor and was also a staple diet of the Wa Swahili and other coastal tribes. The coastal natives were born fishermen and were equipped with all the necessary tackle. All we had to do was to organize them so that they would put their natural ability at our disposal, and our future it seemed would be assured.

The idea, with its hint of adventure, appealed to me strongly and the money was not to be despised either. Lofty guaranteed me a minimum of forty pounds a month to start, a substantial sum in those days, and I had visions of stowing away a small fortune for I knew the cost of living could not be very high. But I overlooked one thing: the need for a written contract. However, a gentleman's word was as good as his bond, or so I had always believed. For good or ill I accepted on the spot and the next morning as I watched the old freighter, silvery in the strong sunlight, dwindling rapidly to the size of a dragonfly, my confidence was considerable and my spirits high. As soon as the aircraft disappeared from sight above the far distant hills I began to prepare for my journey. I was to spend a night or so with Lofty at his house at the foot of the N'Gong hills just outside Nairobi to receive my briefing, and thence travel by train to Mombasa, where I would change to the local bus for the final two hundred miles to the coastal village of Kiwayu. I would then board an Arab dhow for the narrow crossing to the island. Meanwhile I had to travel to Nairobi at least seven hundred dusty, bumpy and uncomfortable miles to the north. Not only did I have myself to think about, but I was responsible for the welfare of a large, self-willed though basically friendly Bateleur eagle, also Rupert who was wonderfully well-behaved and cooperative, and of course Cressida, whom I looked upon as the talisman on whom the success of the whole operation depended. The thought of the journey appalled me, but we had weathered worse storms in the past so no doubt we should survive this little jaunt.

When the next freighter arrived the following week, the captain, Cliff Sowerby, handed me an envelope that contained a note from Lofty and, more important, a number of nice crackly East African pound notes to pay for my fare. The East African equivalent of the Western stage coach was a weekly bus service, owned by an Indian firm, the buses being of the type I have described earlier. They left from the White Horse Hotel in Iringa and took two and a half days for the trip. I had acquired a large light basket of the type native women use to carry their chickens to market, and after surprisingly little trouble managed to ensconce Torquil therein. The basket was loosely woven out of raffia and between the gaps I could see him preening himself happily. Rupert wore, for the first time in his young life, a brand new collar and lead which he accepted at once. Such luggage as I considered necessary was tied to the roof together with the multicolored blankets of the African passengers. As the only European I occupied the front seat with Rupert, grinning happily, beside me. Torquil in his basket took up most of the floor space at my feet and Cressida, now furnished with jesses, leash and hood in case of emergency, sat either on my fist or knee as she felt inclined.

The heat was intense and after a journey compared to which my experiences in a cattle truck as a "guest" of the Third Reich were like a Cook's tour, we arrived in Dodoma where we were to spend the night. Here I was refused admission to the only European hotel in the place. This was not because I was carrying a hawk on the fist, nor because I had an eagle in a basket, but because I was accompanied by a dog! The vaunted British animal-loving veneer at times wears a bit thin in the tropics. Eventually I found I had a choice between sleeping in the open under a wild fig tree on the outskirts of the town, or of accepting the dubious shelter of a crowded and none too salubrious Indian-owned hotel. Thinking of Rupert and the possible proximity of marauding leopards I picked the pub, a bad choice. Under the unfriendly eye of the proprietor-cum-receptionist, I found myself in a small shabby room with two beds, one of which was already occupied by a European of sorts, whose heavy breathing did little to improve my spirits. I unpacked Torquil from his basket and hoisted him on to my left fist and put Cressida higher up on the same arm; the two birds took little notice of each other by now and could safely be left alone together. With Rupert trotting ahead I went out to get some much-needed fresh air and to give the birds a chance to exercise their wings.

The streets were dimly lit as I returned to the huge fig tree to sit and think things out. Metallic chirps, such as could have been made by a swarm of giant crickets, told me that the fruit bats were already at work among the ripe figs. I sat with my back against the tree and fastened the two birds by their leashes to a decaying branch. Rupert pressed close to me and looked up wistfully into my face. For the first time since leaving England I felt homesick, depressed and lonely. Rupert, sensing my mood, jumped onto my lap and, putting his forepaws on my shoulders, gently nibbled my nose in the funny confidential way he had. In the semidarkness Cressida roused her feathers loudly, a familiar well-loved sound. "Damn it all," I thought, "I am not really so friendless or alone." And besides, those sharks were just waiting to be caught.

I returned to the hotel and, lying on the bed in the humid darkness, I ate two bars of chocolate, shared some biscuits with Rupert, and we both drank some tepid water from my old army water bottle. My roommate was still snoring and muttering in his sleep when I must have dozed off. Suddenly I woke, feeling as if an alligator had bitten me. I felt under my pillow for my torch and switched it on. The wall beside my bed was alive with bugs who during my brief sleep had attacked me mercilessly. I grabbed the birds, put Torquil in his basket, called Rupert and fled back to the safety and peace of the wild fig tree. Rupert would have to take his chance with the leopards. I would have fought with the biggest leopard in Tanganyika single-handed rather than face that revolting hotel again. I lit a small fire to keep the mosquitos at bay and smoked a couple of Crownbird cigarettes, the noxious odor of which I should think would have proved fatal to the most determined mosquito. Eventually I lay down between two huge tree roots that sheltered me from the gentle night wind and so at last slept with Rupert cuddled, warm and comforting, beside me. At first light I awoke, cold, stiff and in an evil temper, the welts on my arms and legs where the bugs had dined itching beyond endurance. I returned to the hotel, showed the owner my arms, and after considerable acrimony left in search of a bottle of Milton with which to assuage the fiery torment. Seldom have I been more relieved to see the last of a place than when I said farewell to Dodoma.

The next leg of the journey, unpleasant as it was, was at least bearable. I padded a corner between the window and seat with my bush hat, and dozed most of the way, arousing to some semblance of alertness when we pulled in to a dusty little town where I was able to get a meal of a sort and, more important, to exercise the dog and birds.

The second night's stop was at Arusha, a small compact town in Northern Tanganyika, as pretty as its name, nestling close to the foot of the forested heights of Mount Meru. Here we spent the night at the Meru Hotel, an establishment strongly reminiscent of a Western saloon under the management of a Mrs. Alexander, a kindly woman of uncertain nationality who raised no objection to extending her hospitality to my unusual traveling companions. We ate and slept well, and by morning were ready for anything that fate might have in store for us. On leaving Arusha we also left most of the dirt, dust and corrugations behind us; from now on the road surface steadily improved until, after crossing the Masai district of Northern Tanganyika and Southern Kenya, we eventually got on to a proper metaled road. We then sped across the Arthi Plain with its herds of zebra, family parties of giraffe and the ubiquitous Thomson's gazelle, known to all East African residents as the Tommy, and so came at last to Nairobi, capital of Kenya and the best-known town in East Africa.

At the bus station I looked around hopefully for Lofty. He wasn't there! After considerable difficulty I managed to ring up his home and found that they had not been expecting me until the following day but they would come as soon as possible to rescue me. Meanwhile I dumped my luggage in the bus station under the watchful eye of an *askari*, or native police constable, and went out to take a look at the town, which consisted of an odd assortment of modern buildings, Arab mosques, Indian bazaars and smart European hotels. On the outskirts of the town were acres of the sort of shanty town I had always associated in my mind with Johannesburg. However, interesting as I found Nairobi, and bewildering after months of semi-solitude in the bush, Rupert's reactions interested me more.

Here was a young dog, descended from generation after generation of wild or semiwild ancestors, an animal very close to nature; and yet here he was trotting along the pavements, slipping in and out among the parked vehicles as if he had been city-born and bred.

He followed me with his bushy stern held proudly aloft and waving gently, and eyes, ears and nose alert for any kind of danger or adventure. After losing our way several times in the crowded streets we eventually managed to return to the station in time to see Lofty Whitehead looming above the bustling throng like a bull giraffe above a herd of wildebeests and antelope. Having piled everything I owned into the back of his car and climbed into the passenger seat, with Rupert perched happily on my knee, we set off at terrifying speed for Lofty's home on the outskirts of the city. It proved to be the tropical equivalent of the kind of place owned by business tycoons in Britain and, had it not been for the abundance of black and white fiscal shrikes and the more subtly tinted bulbuls, it might almost have stood on the outskirts of Guildford or Leatherhead. After meeting Lofty's wife and two little girls I was introduced to their bush baby and mongoose (it later transpired that many of the British residents kept these animals as pets). I then fastened Torquil to a block in a shady corner of the lawn and settled Cressida in a well-lighted shed on her own where she had plenty of room to bathe, feed and recover from the journey.

After tea I went into the garden to look at Torquil and found to my horror that he had snapped his leash and disappeared without trace. Luckily the leash had broken within a few inches of the jesses so there was little danger of his getting entangled; and as it was not connected by a swivel but merely slipped through the slits in the jesses there was every possibility that it would work free on its own. I peered frantically at the nearest treetops and saw nothing, though every cluster of dead leaves and deserted crow's nest took on the shape of an eagle and made me sick with disappointment and frustration. Suddenly all the small birds in the neighborhood started shouting abuse and I hurried to the spot from whence it came. There, on the very top of an eucalyptus tree which was bowing with his weight, sat Torquil, his wings half-open to help his balance.

I hastened back to the house for his lure and, calling him by name, flung it to the ground in an open space where he could see it. A Bateleur, because of its disproportionate wing length and shortness of tail to the size of its body, has difficulty in maneuvering in a confined space. Torquil saw the lure, was hungry, and dropped out of the tree; but his terrific speed and lack of experience were against him. He shot over my head, gained height and disappeared in a great arc above the distant roof tops. This performance, though impressive to watch, worried me. The Bateleur is immensely swift on the wing and, not knowing the area, Torquil could well be miles away in a matter of minutes. What really upset me was that he was still a very young bird and most unlikely to be able to cater for himself. I raced back to the house and told Lofty what had happened. He showed little enthusiasm at the thought of leaving his comfortable armchair, but got out the car and we set off in the direction in which I hoped Torquil had traveled. Soon the houses fell away behind us and the semiopen tree-dotted country, that bordered the western edges of Nairobi, began. I scanned the sky, the treetops and even the ground ahead in the hopes that a curious instinct I possess that has many times helped me to find lost birds was at work. Without any conscious knowledge of a hawk's whereabouts I seem to be guided in the right direction. So it was that I suddenly saw the wide-winged, tail-less, almost prehistoric silhouette of Torquil, black against the evening sky. With his jesses streaming behind him he shot overhead and landed in a stunted thorn tree a few

Shark Hunting on Lamu

hundred yards from the road. Climbing up, I forced my way between the bayonet-thorned lower branches, and grabbed his jesses. He clambered happily onto my fist, and I returned to the car, scratched, with torn shirt and shorts, but triumphant. That night Torquil slept with Cressida in her shed.

The following evening I caught the train for Mombasa. On the journey to the coast I met Juni Carberry, one of the last of the "Happy Valley Set" and at that time the owner, with her husband, of the Eden Roc Hotel in Malindi. We were the only travelers accompanied by dogs; the dogs made friends and so did we, and our compartment in no time began to look like a traveling menagerie. Juni asked me to stay as a guest at the Eden Roc and, as she had a hired car awaiting her arrival in Mombasa, we decided to travel together. Somehow we all managed to struggle aboard, but what the Swahili driver thought of his remarkable assortment of passengers goes unrecorded.

Malindi is many miles along the coast to the north of Mombasa and the journey was an exciting new experience for me. There are several creeks that have to be crossed by ferry, the ferrymen chanting Swahili songs as they hauled on the ropes to pull us across. The journey was not uneventful. We reached a broad tidal creek and seeing that the ferry was on the far side decided to give the dogs a run while awaiting its leisurely arrival. I also thought that the birds would enjoy a chance to stretch their wings. Cressida flew to the roof of a dilapidated fisherman's hut, and when called put in a few brilliant stoops to the lure. When Torquil's turn came he suddenly lunged forward, jerking his jesses from my hand and went flapping and gliding across the whole expanse of dark oily water to land on a pile of unsavory-looking jetsam among the mangrove trees on the far side. We hired a decrepit rowboat from an Arab fisherman who had watched the whole performance with some glee, and rowed ignominiously across to retrieve him. Torquil always was a strong-willed and unpredictable bird with a propensity for letting me down at the most embarrassing moments.

Malindi was all and more than a tropical paradise should be. Palm trees waved and nodded their fronds, humming to the tune of the gentle ocean breeze; miles of silver sand were lapped by the almost too warm, blue-green Indian Ocean; further out coral reefs guarded swimmers from marauding sharks, and the only fish were tiny, butterfly-bright, friendly and inquisitive, coming close to inspect the strange human creatures who dared to invade their warm, wet and salty habitat. The Eden Roc Hotel itself might almost have been transplanted from the French Riviera, and my fellow guests, being of the type one expects to find lazing on some sparkling Mediterranean beach helped the illusion. Where the shore and bush united stood huge wild fig and banana trees and as I walked there with Rupert before turning in, I could hear once more the familiar clinking cries of the fruit bats, and see in the moonlight their wide-winged forms circling like huge moths as they searched for the most tempting of the overripe fruit.

The next morning when I went to catch the bus that was to take us on the last lap of the journey to Lamu I had a nasty shock. The one bus of the day was Arab-owned and driven, and most of the passengers were either pure Arabs or Bajunes, people of mixed blood in which the Arab element predominates. Whatever their racial origins they all had one thing in common: they were fanatical Moslems. To them all dogs were anathema, unclean beasts, to be despised and avoided and here I was about to spring happily into the front seat with Rupert under my arm. A violent argument began; my Ki Swahili was strictly of the kitchen variety and I had to enlist the help of Juni Carberry as interpreter. It appeared that the alternatives were either to put Rupert in a box, lashed to the roof for the whole journey of about a hundred miles, which was unthinkable, or to leave him behind which was out of the question. It seemed that the only solution was to hire the whole bus, leaving the passengers to travel on the next one twenty-four hours later.

Eventually, after much argument, we reached a compromise. I was to hire the front half of the bus while the other passengers, of which there were luckily not many, would cram together in the rear. This cost me a pretty penny and I eventually arrived in Lamu with just fifteen African shillings between myself and destitution. This, however, worried me not an iota because I was full of confidence as to the vast fortune I was about to reap. Little did I realize at the time that I was not to receive a cent and that the only thing I would gain was a wealth of experience and a certain cynicism regarding potential employers that took a considerable time to wear off. The journey to Lamu was, as might be expected, as uncomfortable as its predecessors, and the surly glances from my fellow passengers did little to cheer me on my way. In addition to their ingrained fear and hatred of dogs, they did not think much of Cressida and Torquil either. After the cheery good humor of the up-country Bantu tribes I found this attitude singularly unpleasant. It was to continue in varying degrees throughout my entire stay on Lamu.

On arrival at the only hotel on the island, an establishment owned by a gentleman known as "Coconut" Charlie Petley, being tired and filthy after the journey, I asked if I could have a bath. He rather grumpily acquiesced, but asked me to wait until he removed the beer! It appeared that, lacking a refrigerator, he kept his bottles of beer in the bath, immersed in tepid sea water, this being the only liquid available other than the alcoholic variety. As I was the only guest I think he resented the trouble entailed, and as for Rupert and the two birds he made it abundantly clear what he felt about them.

The next day I met Alan Bidder and moved in with him. Alan, who came originally from Henley-on-Thames, was living in an Arab-type house on the extreme eastern edge of the island. He was a splendid character, larger than life in personality as well as physique. He must have weighed about two hundred and fifty pounds and was bone idle with it. How he came to be in Lamu escapes me, but there he was, happily insolvent and completely without thought for the morrow. As neither of us had any money we managed to persuade a friendly and helpful Indian shopkeeper to let us run up enough credit to live on. I was confident of making a fortune out of the sharks and crawfish, and Alan was equally enthusiastic about helping me to spend it.

Lamu island may appear romantic in an Arabian Nights sort of way to casual visitors down from Nairobi for the weekend, but Rider Haggard, who stayed there when writing *Alan Quartermain*, described it well, and it had not changed much when I descended upon it sixty years or so later. Filth, flies and smells, he would have remembered them all, not to mention the hordes of apparently ownerless cats that prowled the night-shrouded foreshore in search of dead or dying fish. To a naturalist, Lamu had one redeeming feature—the innumerable old buildings, ruins and catacombs were the homes of myriad bats of several species. There were horseshoe bats that seemed almost identical with the greater and lesser horseshoes indigenous to southern and western Britain, and there were a number of small pipistrelle-like insect-eaters. One particularly jolly beast was quite large, bigger than a noctule, with silky gray fur, astonishing pure white wings, a head rather like a prick-eared terrier and, as I painfully discovered on my first close acquaintance with one of these intriguing features, jaws of incredible power. They lived singly or in pairs under the eaves of the Arab houses and in alcoves of the administrative offices, and seemed entirely at home in full daylight. They could run and climb like squirrels and their flight was dashing and falcon-like as they hurtled about the sky with pointed wings flashing snow-white in the glare of the sun.

A few days after my arrival I hired an ocean-going Arab dhow called the *Skandavia*. A sinister-looking vessel, I could picture her under full sail, her holds crammed with African slaves as she drove before the monsoon on her way to the Persian Gulf. Lamu once had had a notorious reputation as a slaving port. The crew of Bajunes were pleasant enough fellows but most suffered from elephantiasis. The first time I put to sea with this villainous-looking crew we anchored for the night in a bay on the neighboring island of Manda. To escape the attentions of the sand flies I moved a few hundred yards inland and lay down under a large shady tree. Tired out with so much sea air and hot sun I slept the sleep of the proverbial just beneath a single threadbare blanket. After what seemed a few minutes I was jerked awake by the voice of Rupert who had slept beside me and was now barking frantically. As I raised my head a few inches to see what the fuss was all about I saw in the half light of early dawn a pair of enormous pachydermous columns within a few feet of my recumbent form. "My God," I thought, "this is it! I am about to be trampled into oblivion, and I did not even know there were any elephants on the island." Glancing upward, I was still only half aware of the horror of it all when, as my eyes began to focus, I saw that the legs disappeared into a dirty loin cloth. Relief swept over me; it was only the *nakodah*, or captain, of the good ship *Skandavia*, who had brought me my early morning cup of tea.

This *nakodah*, despite his rather unprepossessing appearance and physical misfortune, was a most genial rogue with a pronounced, if wry, sense of humor. Once, when out with the dinghy, "Baby Skandavia," three members of the crew, including the cook, were decanted into the sea by a sudden squall. The captain bellowed to the rescue party, who were about to throw out lifelines, *kamata quanza m'pishi* (grab the cook first). He was also somewhat broadminded in his religious beliefs and, though probably as good a Mohammedan as the majority of his fellows, raised no objection to carrying Rupert about on all our voyages. Rupert, for his part, learned to curl up out of the way; indeed he soon became an extremely competent and self-confident sea dog, and would leap from shore to ship or vice versa under pretty trying conditions. From early puppyhood he had proved to be highly adaptable, a wonderful companion and guard, and it was a pleasure to take him about with me. Here, on this Mohammedan stronghold, where he was the only dog on the island, it was imperative that he should behave in an exemplary fashion. He was there on sufferance and he never let me down.

I soon made my headquarters on the island of Manda, which was separated from the larger Lamu by a narrow channel controlled by savage currents. To fall overboard would in all likelihood have had disastrous results. Manda was supposed to be the best place for crawfish, which lurked in caverns in the coral reef, and for sharks, which cruised about the open sea like piratical submarines, ever on the alert for prey. The method of catching them, as used by the local fishermen, was simple if unsporting. A series of empty petrol cans, or *debis*, were attached at intervals to a length of powerful line from which other lines, each with a vicious-looking hook baited with rotten fish, were fastened. The petrol cans acted as floats and the whole contraption was allowed to drift about on the calm surface of the sea. The sharks, attracted by the stench of the bait, hooked themselves and then set off towing the lot—lines, cans and all—until exhausted. The cans of course returned to the surface after the shark had shot its bolt and acted as markers for the pursuing fishing boat. The crew had only to haul in the short line to which the shark was fastened and slay the brute with clubs and pangas. Very little risk was entailed. In many cases, however, all that the hopeful shark fishermen might find would be the head, tail and backbone—other sharks, and the even more ferocious barracuda, would see to that.

Catching crawfish was much more fun. These huge beasts were as big or bigger than the largest lobsters, which they superficially resembled. They lacked the lobster's great crushing claws, but made up for that by having enormously elongated antennae, which protruded like wireless aerials and betrayed their owners' presence as they lurked in their cavities in the coral. These antennae were often the only indication of the crawfishes' presence and gave them their Swahili name of *kamba* meaning "string." In those days goggle-fishing, with all its paraphernalia, frogmen's suits and snorkel tubes, had yet to become the popular sport it is today. All I had in the way of equipment was a contraption of glass mounted in a wooden frame, made for me by a local fisherman friend, which gave me some sort of aid to underwater vision, a trident of the sort favored by Father Neptune, and my own lungs, which must have been sorely tried at times.

I would take a deep breath, slip gingerly into the warm, transparent water and climb about the faces of the coral cliffs looking for the betraying presence of the crawfishes' antennae, which after a time I became quite adept at spotting. If I found a crawfish I would lower a piece of red flannel or, better still, of putrid fish, on the end of a line and let it hang tantalizingly just outside the crawfish's refuge. If the crustacean was hungry, or merely inquisitive, it would slowly emerge and clamber spider-like among the waving weed and razor-sharp projections of coral. When I estimated that the beast was far enough from its refuge I would submerge rapidly and try to prong the unfortunate with my trident or else grab it before it could hurtle backward into its den.

Catching crawfish by hand had its disadvantages as the animals were covered with tiny excrescences, sharp as broken glass; these could, if carelessly handled, cause some very nasty sores which, perhaps due to the chemical qualities of the sea in that part of the world, took a long time to heal. Some of the caves were also the homes of giant moray eels which, according to native hearsay, were not only deadly poisonous but had a reputation for being able to bite clean through a man's thigh. This may not have been complete exaggeration because once, in a moment of bravado, I speared a smallish specimen and it immediately turned and snapped the spear-shaft clean in two!

All the crawfish I caught were kept alive in a submarine compound, made of sections of expanded metal lashed together, and here they appeared to thrive until the time came to fly them up-country to their final destination.

This may appear to have been an ideal life, the classical lotus-eater's paradise, but in fact it had its disadvantages, in my case the greatest being that I never received any money. Even my Indian storekeeper friend was becoming a little less affable. The only one of my immediate circle who appeared to revel in the life was Torquil, who would go racing through the sky above Lamu or Manda, rocketing about at fantastic speed, his wings tipping and tilting like a tightrope walker's pole. Since he was rarely hungry, he must have been providing for himself with remarkable efficiency. Both Rupert and Cressida (who must have been at least eight years old by this time) found the humid, enervating climate too much for them and both had become listless, disinterested parodies of what they had been. Rupert, at dawn or late in the evening, would sometimes shake off his sloth sufficiently to enjoy a short walk with me along the beach, where he would have a great time chasing the land crabs who always managed to beat him to their burrows above the tide line.

Both Alan and I had decided it was time to leave Lamu. I was getting a small monthly allowance from home and so could, I suppose, have described myself as one of the last of the remittance men, a breed now probably as rare as the okapi. After considerable acrimony I managed to persuade my employer in Nairobi to stump up fifteen pounds, and with two months' accumulated allowance transferred from Barclays Bank in Nairobi I felt like Croesus. Anyway I was able to settle my account with Alibhi Muftizadi and buy two bus tickets for Malindi. We had the same old rigmarole over Rupert as on the previous journey, except that this time Alan and I, with Rupert and the birds, occupied the front seats, and there were two of us to weather the hostile stares and uncomplimentary asides. Thus ended the saga of Lamu and the abortive fishing industry, but at least I had learned one thing—never, under any circumstances, do anything without a written contract. Only thus can one hope to sidestep the human predators rampant in an unjust world.

5

Four Become Three

AND SO WE CAME ONCE MORE to Malindi as dusk was deepening over the coastal bush country and the nightjars were calling from the forest fringes that skirted the long silver ribbon of road that ran toward Mombasa. Dog-tired, desert-dry, corrugated with dust and sweat, we sought refuge in the Sinbad Hotel. Longing for sleep and sustenance, lugging a basket with Torquil in it, with Cressida on my fist and Rupert trotting beside me, we approached the reception desk. "Well, well, well, look who's here," said a voice I vaguely recalled. I spun around. There before me, immaculate in crisply clean bush jacket and khaki shorts, stood my old schoolmate and companion from the prison camp, Peter Mumford and, wonder of wonders, he owned the Sinbad, every stone of it. Greetings and introductions over, he took us to our rooms. Later on, bathed, shaved and generally rehabilitated we trooped into the bar, to be welcomed by Peter, his wife Pauline, father Philip, and stepmother Marjorie.

It was an evening to remember. Never was ice-cold lager more refreshing, never did fresh pawpaw and deep-sea fish straight from the Indian Ocean taste better. Even the slightly

Four Become Three

accusing look of the crawfish reclining on a sea of salad did little to deter me. It appeared that Peter had married Pauline a year after his return from the POW camp at Brunswick, which was where I had last seen him, and that the whole family had migrated *en masse* to Kenya. Here they had organized the building of the hotel, which in comfort and everything else equaled if it did not actually exceed its nearest rival, the Eden Roc. One feature particularly intrigued me: the dining room was open at each end to allow the fresh breezes from the ocean to blow through and thus keep the temperature as near perfect as could be. One end opened practically onto the beach and I was fascinated by the marauding bands of bright coral-pink land crabs that would come scuttling into the room to purloin crumbs and other morsels of food that might fall from the tables. Inches from one's feet they would help themselves to anything that was going with commendable effrontery, indifferent to the interest they were causing. I can, however, imagine that they might have induced anyone with less pronounced zoological interests to take the pledge on the spot.

Other potential allies to the temperance movement were the mud skippers, intriguing small fish with protruding frog-like eyes and pectoral fins which their owners used like feet. Although largely aquatic they had the disconcerting habit of emerging from the sea to bask on the sands over which they would go bounding like miniature sea lions. Not content with this, they would shin up the stems of the mangrove trees to sunbathe on some convenient projection. The first time I encountered these amazing amphibious fish I all but took the pledge myself.

The first night at the Sinbad was one of the best of my life. With Torquil and Cressida happily ensconced in a luxurious bathroom, where they slept perched on the rim of the bath so that their mutes could be liquidated at the turn of a tap, and with Rupert snoring the deep regular snores of utter contentment curled on a mat beside the bed, I slept as I have rarely slept before or since, and rose as the sun was climbing out of the sea to take myself and my three companions for a dawn patrol. Alan slept vociferously on; no early morning gallivanting for him if he could help it. We were alone on the beach as I cast Torquil off. The sight of his falcon-shaped wings, close to six feet from tip to tip as he rose almost vertically playing with the breeze as he shot out over the waves, was breathtaking. Returning in a great semicircle he dropped almost to ground level and came streaking along the beach, his very lack of altitude emphasizing the speed at which he was traveling. Reaching me he climbed to twice treetop height, almost standing on his tail with the abruptness of his ascent, and looping the loop came down in a shattering stoop to land light as a ballet dancer beside the expected lure. Old Torquil seldom caught anything, but nonetheless he was a flying machine without peer when once he made up his mind really to put the power on.

After breakfast Philip and Marjorie Mumford asked me if I would like to join a small party of their friends who were going to the Tana River on a crocodile-spotting trip. I put this as a suggestion to Alan, still stretched like a singularly well-fed oriental potentate on his bed. "Crocodiles," quoth he, "who the heck wants to see crocodiles? After all, old boy, what are they? Only things which open their mouths and get turned into shoes."

Alan and I decided to prolong our stay in Malindi for a few days. There were several reasons for this, not the least of which being that we had neither of us the faintest idea of what our future was to be. Above all we had been invited to stay on as the guests of the Mumfords, and VIP guests at that. On the second night as I returned to my room, glowing with sunburn and good wine, I flopped into bed like a contented walrus and a pain like that caused by a hundred red hot needles shot through the big toe of my left foot. This is where I came in I thought, as I sat cursing, trying to staunch the flow of blood with a bath towel, and my thoughts flew back to Lhoni, the irascible but endearing mongoose that had ruled my grandmother's home thirty years before. I was right: curled in a tight reddish-gray ball at the foot of my bed, leering at me from insolent bird-bright eyes, lay Mango, or it might have been Marmaduke or even Mitsi.

Whereas some exceptionally devoted animal people may be sufficiently ill-advised and long-suffering to keep one mongoose, the Mumfords kept them in droves. They were not of course the same species as Lhoni; they were tiny, almost weasel-sized, sociable pygmy mongooses, which live in little troops all over the drier parts of East Africa, often making their dens in the labyrinthine tunnels of a termite mound. They make splendid pets provided they happen to like you and they have all the comical ways and cheerful impertinence of the better-known Indian gray mongoose. Called in Swahili, *kitate*, they were often offered for sale for a few shillings by the coastal tribe who were quick to spot and exploit the European weakness for adopting pets of this sort. I never discovered exactly how many mongooses the Mumfords were owned by, but they seemed to be everywhere and, being as good vermin deterrents as their Indian counterparts, must have been of enormous help in keeping down the snakes, rats and other unpopular fauna. Living under conditions of complete freedom and with as much to eat as they required, they bred freely, and I was entranced to watch one of the females, Mitsi I think, playing with four fat little miniatures of herself. They were leaping all over her, growling, seizing her tail or any other part of her that came to mouth and worrying it with their tiny needle-sharp teeth; they would roll on their backs with their feet in the air as they juggled with an enormous seashell, looking rather like a litter of undersized dachshund puppies but with a certain elusive enchantment that was theirs and theirs alone.

We all enjoyed our respite here at Malindi, and Cressida, who was at last showing inevitable signs of age, was exceptionally happy. She would spend all day on a sunny balcony with a dish of fresh water for drinking and bathing and the security of the shady cool bathroom to retire to if she wished. She would still come for a stroll on the cooler evenings when the cicadas had begun to tune up their instruments and the first bats were flitting from their diurnal roosts in crevices in the old Arab houses. She would follow me, flying from palm tree to palm tree, occasionally dropping to the ground to seize and devour some tiny crab or beetle; her appetite was, as always, prodigious. She had developed a sudden antipathy for Torquil, and had not the slightest compunction about attacking him. She would suddenly land on his back, even his head, gripping hard with her small powerful talons. He, for his part, bore her no ill-will but nonetheless, having seen what he could do with his enormously powerful beak when he meant business and having more than once felt the grip of his feet when he was in a temper, I decided to keep them separated.

Splendid as was this life of eating, drinking and loafing in the sun or sea, the migratory urge was stirring, and I decided it was time to be off to seek what surprises life might yet have in store for us. Rupert, Cressida, Torquil and I were a united team, a family almost. To those who have never had this kind of relationship with animals it is difficult to explain the curious interdependence. I loved my dog, my eagle and my kestrel all in their different ways, and I had every reason to believe that they felt the same way about me. We had a final and uproarious evening in the Sinbad, as a result of which Alan, sorely stricken, decided he was unable to travel unless his bed went with him; and so, having said our *au revoir*s, my companions and I came

Four Become Three

by devious methods of transport to Nairobi, and thence found ourselves at the Mananga River Hotel, close to the Tanganyika border, and on the boundary of what was to become the famous Amboseli National Park. The hotel was then owned by one "Budge" Gethin, a reformed white hunter, who also maintained a small hutted camp in a clearing deep in the surrounding acacia and thorn bush country some miles from the hotel. Known as Rhino Camp, this later became the headquarters of the national park. It was one of the best places in the whole of Africa for watching some of the more exciting larger game animals, as they led their own undisturbed and unaggressive lives, watched over and guarded by the great round snow-sprinkled bulk of Kilimanjaro some miles to the southeast.

Not far from the hotel I found a colony of big black-bodied white-headed buffalo weavers; they were busy about their nests, weird structures shaped like distorted water bottles, with the entrance in the neck so placed that one wondered how the owners managed to enter without the greatest difficulty. I enjoyed watching the weavers, cheerful noisy extroverts, chattering and calling like a group of cockney housewives, and busy searching the brick-red ground for scattered seeds. However my interest in them vanished when I noticed, perched at the top of an acacia tree in the middle of the weaver colony, what I at first took for a shrike but which on closer inspection proved to be a minute shrike-sized hawk. In fact it looked like a diminutive, pearl-gray-and-maroon kestrel. As I watched, the bird dropped onto a large locust-like grasshopper, tore off its head and wingcases and, carrying the remains in one tiny taloned foot, swung upward and with hardly a pause vanished into the nozzle of one of the weaver's nests. Judging by a modified rendering of the sound made by a family of peregrine eyasses, the nest was obviously occupied by a brood of these fantastic little birds. Later, on consulting my copy of Austin Roberts' *Birds of South Africa*, I discovered that what I had seen was a female pygmy falcon, one of the smallest and certainly one of the most enchanting raptors in the world. I found also that these little hawks were quite common in the dry thorn country. With their predatory instincts scaled down to suit their size, they must have done a great deal of good in keeping down locusts and other unpleasant insects.

High up over the hotel, above the forest canopy, loomed the great bare head of the mountain, Oldonyoroc. I used to watch the purple dusk come down wrapping the heights in a mantle of mystery. Helping in the hotel was a girl, Nancy Russell-Smith. The last evening I was to spend there that visit, we talked about the mountain and made a plan. One day I would come back and we would climb it together; we would build a cairn of stones on the top to prove to ourselves and to any who might follow that had been there. Months later we fulfilled our ambition. We did climb the mountain, built our cairn, and hidden in a crevice at the very top is still the tin perhaps, with our names and the date, which we placed there so carefully.

Rupert, Cressida, Torquil and I left the Mananga River Hotel and came once more to Arusha, the capital of the Northern Province of Tanganyika. I was feeling far from well and I wanted somewhere to take refuge and plan our next move. The Meru Hotel accepted us, if without great enthusiasm at least without undue fuss. This hotel, like so many in East Africa, consisted of a central courtyard, cool and shady, in the center of which grew a dense but rather stunted fig tree. The courtyard was surrounded by rondavels and bungalow-type huts, the main block being some distance away, consisting of reception office, dining room, kitchen, etc. I was given a key to a rondavel and retired there, after first settling Torquil, now jessed and leashed, on a low branch of the fig tree, where he caused a lot of interest and provoked some pretty fatuous questions from some of my fellow guests.

I took a jug of lime and ice water and retired to my quarters to await the onslaught of fever that I knew to be approaching. In the small hours I was smitten by what seemed to be the result of a hundred simultaneous attacks of flu magnified beyond belief. I had read often of malaria, but now here it was in all its unabated horror. At one moment I was burning with all the fires of hell, the next freezing as if plunged into some icy mountain torrent, the while my bed shook and reverberated as if rocked by some demon hand. My mind was tormented by unspeakable visions, and I lay longing for final release. Then, during a sudden withdrawal of the onslaught, I saw Cressida perched rock-like on my bedrail. As I dragged myself over into what I hoped would prove a slightly less uncomfortable position, I glanced out of the open doorway of my hut toward the wind-tossed fig tree; there I saw the big black bulk of Torquil, motionless and serene, as he rode the swaying branch on which he sat. Deep regular breathing from the mat beside my bed told me that Rupert was sleeping, happy in his master's nearness, and a patch of moonlight showed where he lay tightly curled, colored almost like an agate in the unreal light. These three were confident of the future and I knew, even as another wave of fever gripped me, that I would pull through. I finally fell asleep, to wake late the next morning soaked with cold sweat and as weak as a newly whelped puppy. The fever had left me, not to return, until many months later.

For the next few days I felt much as Samson must have felt after his forced visit to the barber, and spent most of the time lying in a deckchair in the central court of the hotel with Rupert lying at my feet (when he wasn't absent on mole-rat hunting expeditions). I drank phenomenal quantities of fresh lime juice, ate pounds of mangoes and pawpaws and pondered deeply on my future. I also laid into enough Paladrin, a malaria-prevention pill, to save an infantry battalion, and spent much time chatting with other guests.

There are few places on the face of this earth where one is

more likely to run into "characters" than in an East African hotel, particularly in the smaller, less ritzy ones. There was, for instance, Cleland Scott who kept lions as members of his family long before it became the thing to do. There was Ewart Grogan, who at the turn of the century walked from Cape Town to Cairo, negotiating some of the most dangerous and unpleasant country in the world, because the girl he wanted to marry refused to accept him until he had proved himself. He took, I believe, two years to complete the journey, but the girl waited and when I met him they were still happily married after fifty years.

One evening, as I was returning with Rupert after exploring the foothills of Mount Meru, an enormous Norwegian strode into the bar and bellowed for a bottle of Tusker. I was also thirsty; the bar was empty and we began to talk. Jan Kielland was one of a large family of Norwegians who owned a coffee *shamba*, carved out of the primeval rain forest at Oldeani, not far from the Ngoro-Ngoro crater. He was in Arusha to collect a spare part for his jeep. We talked about butterflies—he collected lepidoptera for the Oslo museum—and we discussed the future exploration of little known places. We talked for hours about big game and small game, wild animals and tame, and we drank much Tusker in the process. Finally we retired to bed as the great backdrop of Mount Meru's forested slopes echoed with the good-night cries of home-going bush babies and the maniacal screams of awakening tree hyraxes.

The result of all this was that later that day I found myself, with Rupert, Cressida and Torquil, bucketing over the corrugations as the yellow jeep, with a wild Viking at the wheel, ate up the long and dusty miles at fantastic speed on the way to the Kielland farm, Kit'Ndovu, in Oldeani, a hundred and fifty miles southwest of Arusha. Oldeani was the real Africa;

here could be found all manner of beasts both known and unknown to science. The *shamba* itself stood in a clearing of the forest that surrounded it on three sides, the fourth looking out over a wide escarpment, which seemed to drop away almost vertically to the blue-gray rhino-haunted scrubland thousands of feet below. Here, standing at the door of my rondavel, which was perched like an eagle's eyrie upon the only few square yards of flat surface in the area, I could watch the augur buzzards and white-necked ravens circling far below me or riding upon currents of warm air that rose from the hot arid lands so far beneath.

Jan Kielland lived with his widowed sister and her two small children in the cool comfortable farmhouse, which arose so surprisingly from the clutches of the encroaching forest. Here they grew coffee and fruit trees, and lower down, where the heat was more intense and the shade trees fewer, both maize and sunflowers. Jan was a hard and conscientious worker but his head, like mine, was as often as not filled with dreams of exploration of the exciting unknown depths of the forest so close to our door, and whence came the loud trumpeting calls of touracos, the pheasant-like calls of francolins, and the machine-like sounds of the elusive crested forest guinea fowl.

As the evening air began to cool I would call Rupert and, with Cressida, who seemed to revel in this strange untamed country, we would sometimes follow the course of the irrigation furrow that arose far in the deep green leafy twilight world of the rain forest. These expeditions were fascinating. Rupert would trot silently ahead, his auburn ruff rising slightly at the scent of danger lurking nearby, and danger there most certainly was; the air was thick with the curious barnyard smell of buffalo. Once I all but tripped over a huge solitary bull, black and shining, with curved horns carunculated and bossed, elemental as the mountainside behind him. He was on his feet in an instant and I looked into the huge, almost soulful black eyes of what is commonly supposed to be the most dangerous animal in the world. All this I saw in the few seconds before he turned and went pounding and crashing away into the thick green creeper-entwined forest wall, which opened to receive him and closed behind him so closely that only the thunder of his retreat and the flattened rank-smelling patch of grass that had been his resting place showed evidence of his existence.

Once, rounding a bend in the narrow forest path, I saw the huge gray bulk of a rhino, stretching from one side of the track to the other. He swung his great prehistoric head in my direction, a foot or so of creeper still clamped between his jaws, his ears flicking and his nostrils twitching as he sought to catch my scent. He knew something was amiss, but his myopic eyes failed to tell him exactly what was the nature of this threat. Suddenly, with his tail erect, he swung his head with its magnificent but terrifying pair of horns and came trotting forward, light-footed as a polo pony. The bank to my left rose almost vertically and I grabbed the branch of a low bush and hauled myself up as now, less than twenty yards away and gathering speed, he began the horrible asthmatical snorting, so characteristic of a charging rhino. He lowered his head and came storming past, so close below me that I could have touched his great mud-encrusted back with my foot. When he had gone by I dropped again onto the track and returned home the way I had come, noting where he had turned off the path and followed one of the old trails, used by his kind since time immemorial.

The irrigation furrow was a natural meeting place for all sorts of exciting birds and butterflies. Among the latter were some magnificent insects whose great angled wings seemed to have been fashioned out of mother-of-pearl, and whose flight, erratic and deceptively speedy, defied my efforts to net them. There were also velvety-black and metallic green swallowtails by the score, and the dashing falcon-flighted chestnut red charaxes, with their disgusting passion for the juices of fresh animal droppings. I used to lie on my tummy for hours and watch the birds bathing and drinking at a favorite water hole; their variety and beauty almost too spectacular to be true. There were the long-tailed, jaunty crested paradise flycatcher, the ethereal blue flycatcher, the rare and furtive forest-dwelling Gurney's orange ground thrush, and the lovely plump lemon dove, with its blend of yellow, green and pink plumage, so conspicuous in the sunlight of the furrow side but so hard to spot among the shadows of the deep rain forest where they lived.

On these solitary expeditions into the forest I used normally to leave Rupert and the birds behind in the house, partly because the dog's presence would alarm the wild creatures I wished to study, but more especially because of the danger of leopards who, lurking unseen on a branch overhead, might have seized and made off with them in a flash. Torquil had a favorite perch in a thick fever tree behind the house, from which, when so inclined, he would launch forth and go rocketing out far over the Korongwa. Surveying the ground from this great height he no doubt kept his weather eye open for any delicacies that might tempt his voracious appetite. He used to go further and further from home, and be away for increasingly longer periods. As often as not he would be joined by one or more wild Bateleurs, who would put in some dramatic half playful stoops at him, to which he would reply with some very pretty aerobatics of his own. However long he was absent and however far he traveled (he must have covered hundreds of miles at his breathtaking speed) dusk would find him sitting hunched, almost owl-like, on his favorite branch close to the trunk of the tree. And then one evening he didn't return. With no warning of his intentions he just failed to turn up and we never saw him again, or if we did he had suddenly gone so wild as to be indistinguishable from the other immature Bateleurs in the area, of which there were many. I was sad to lose him, and missed him a good deal, but I like to think of him sailing freely over the vast blue distances of Africa that were, after all, his birthright. Neither Cressida nor Rupert were noticeably upset at Torquil's departure, and we three became now, if possible, even closer.

6

A Double Loss

I HAD APPLIED for a course at the Egerton School of Agriculture at Njoro in Kenya, had gone for an interview with the principal in Nairobi and had been accepted, and now I had a few months to fill in before the beginning of the next term. The Kiellands and I had become very friendly and we used to go on long butterfly expeditions deep into the rain forest whenever Jan could spare a few hours from running the farm. Margit, too, was interested in all forms of natural history and would sometimes leave the children (Anne-Elise and Per-Erik) with an *ayah*, or native nurse, and come with us on some of our shorter safaris. Margit was indeed a surprising person, small and dark, vivacious, and with a keen sense of humor. The only really Scandinavian characteristic about her was her physical toughness. She could indeed walk me into the ground and I think both she and Jan were curious to see just how much physical stamina I possessed.

So the exploration and descent into the Oldeani crater was planned. We had often sat outside the farmhouse in the evening, drinking coffee and gazing at the crater rim which rose out of its shrouding canopy of bamboo miles to the west of the farm. No one, we believed, had ever been there before us. Mile upon mile of forest intervened, the haunt of elephant, buffalo, rhino and heaven alone knew what other denizens of the African wilds. High overhead, tiny in the limitless blue, swung and circled the great crowned hawk eagle, most powerful of all Africa's innumerable birds of prey. Its wild, musical, ringing call came drifting down to us as we gazed skyward. Here was the very spirit of the rain forest, the self-appointed guardian of the crater, sending forth its challenge to us where we stood thousands of feet below. We made our plans, but we would have to wait a while until the coffee berries, now red and juicy on the dense green bushes, had been picked and harvested.

Cressida had taken over Torquil's old roosting place in the fever tree. Here, sheltered and secure, safe from the fierce heat of the African sun, she would doze away the hours, preening and grooming herself. She still kept herself immaculate. Sometimes she would come down for a dust bath, a cool drink, or a succulent grasshopper or beetle. As late afternoon brought the shadows creeping out of the forest she would become more lively, and she, Rupert and I would go for short walks about the estate, sometimes as far as the neighboring farm, which was occupied by an Afrikaans family who had become great friends of ours. There was a clearing in the forest, not far from the house; here there was a shallow pool where the lovely foxy-red bush buck with their limpid eyes and dagger-sharp horns used to drink at dusk. Here Cressida, Rupert and I would come to watch and to think. Rupert, knowing my moods, seemed as interested in the local fauna as I was, and he knew that all horned game were taboo, if only for his own safety. He was, however, a considerable naturalist in his own right, and was forever astonishing me with the number and variety of small rodents that he brought in, many of which I am sure have never been classified.

A Double Loss

These included an array of attractive dormouse-like beasts, with bushy tails and horizontal black stripes along their red, fawn or chestnut-colored bodies. There was also a colony of small, dusky, almost black animals, with vestigial tails, that closely resembled hamsters but were obviously quite different. These lived in a particular corner where cultivation gave place to wilderness, and to the best of my knowledge, were found nowhere else. We also discovered a peculiar chameleon, clambering among the coffee bushes, which, according to Dr. Leakey of the Coryndon Museum to whom I later presented it, was most certainly a new discovery.

Cressida would paddle about on the fringe of the pool, soak her breast plumage and tail, and retreat to a comfortable sunny stump to relax and dry off. She would sit and gaze at me intently, and I feel sure that in her own way she was thinking back over the years of adventure and companionship that we had shared. One evening when I called to see her where she sat on her chosen perch in the fever tree, she was asleep, her head buried in the feathers of her scapulars; I left her undisturbed to sleep on. Much later I came to carry her indoors to her own roosting place in my rondavel where she used to spend the nights safe from nocturnal predators. Gently I picked her up and she nestled into my palm, holding on with her little yellow, black-taloned feet. I carried her into the circle of light from my pressure lamp. She turned to look straight into my eyes and gave a sudden convulsive jerk, her head lolled sideways, her wings flicked open and she was gone. And with her went nearly ten years of love, companionship and adventure. At first light next morning, with Rupert beside me, I buried her under the fever tree, her own chosen last refuge. Perhaps she is still resting there today. So passed Cressida and a little bit of history.

Rupert and I were now closer than ever, but fate had not quite finished with us. During my stay with the Kiellands I had become interested in coffee planting, and was anxious to buy a farm in the district, if only to enable me to continue my study of the fantastically varied fauna. A certain land unit became available, but it was necessary to apply to numerous authorities, all of whom vied with each other to know what qualifications and reasons I could possibly have for wishing to take over and run such a farm. Also the question of my ability to put down a deposit arose. I decided to go into Arusha, the seat of local administration, and thrash the matter out with all the appropriate authorities. I would be busy for a number of days; the weather was at its hottest and driest, and Rupert, used to the vast freedom of Kit'Ndovu, would be miserable in Arusha. By foulest fate both Jan and Margit were ill with malaria, and the children were under the care of the African *ayah*, a pleasant and competent Wam'Bulu girl who was very fond of them. Our neighbor on one side, a retired ex-gunner major, agreed to look after Rupert for the few days I was to be away. Rupert and the major knew each other, so I assumed all would be well.

I went by bus to Arusha and duly set about dealing with, and where necessary conciliating, authority. I had booked in to my old rondavel at the Meru Hotel and late on the second night, my business completed, was about to retire to bed ready to return to Oldeani the following day when there was a sudden commotion. Into the bar strode the major with whom I had left Rupert, and whom I had believed to be two hundred miles away. He called for a large brandy and had just received the glass when I arrived to demand to know what had happened. The major, it appeared, had stopped for many brandies at different Indian shops along the route and was, to put it mildly,

A Double Loss

a great deal drunker than any lord that I had met up to that time. He blurted out an almost unintelligible story about how Rupert and his bull terrier had started a fight, how he had with difficulty separated them, and how to save a repetition of this sort of thing he had put Rupert in the back of his van and driven off with him, intending to bring him to Arusha to join me. At a small township called Mtu-wa-m'bu (river of mosquitoes) about fifty miles from Oldeani, he had stopped for petrol and a drink. To his horror he had found the rear door wide open and no sign of Rupert. He had, so he told me, driven back over the route, and stopped to call and search at various points but had found nothing. Then, fearing my wrath, he had drunk a considerable amount of brandy and had driven on through the night to tell me the awful news. By the time he reached the Meru Hotel he could hardly talk, let alone drive, so realizing that nothing could be gained by violence, however justified, I looked for help elsewhere.

In the courtyard of the hotel stood Mike, an Englishman whom I knew slightly. I told him my story and without hesitation he agreed to help. After a quick drink apiece to spur us on our way we jumped into his safari truck and drove all the way back to Mtu-Wa-M'bu and up the escarpment, stopping and calling, calling and searching. There was no response; all I could hear was the whooping of spotted hyenas and the distant call of a Verraux's eagle owl. Sick at heart I returned to the van; there was little more that we could do that night. We drove back to Kit'Ndovu where the Kiellands, still weak with fever, gave us coffee and a meal that I couldn't even look at. Next morning at dawn I borrowed Jan's heavy rifle and with only a water bottle and a bar of chocolate in my pocket, asked young Mike, who had helped me as much as he could, to take me where he thought the van door might have worked open. He did so, wished me luck and drove off back to Arusha to attend a farmers' meeting.

So the search began. I walked from early light until dusk, searching every foot of road. I asked parties of Wa Kamba and Wam'Bulu tribesmen if they had seen a small red dog. None had and no one showed the slightest interest. What, after all, was a lost dog to them. That night, sheltered in a small cave at the foot of the escarpment, I drank some water and ate half a bar of chocolate, and I hated my thoughts. I slept but little and the first light of a new day came to me as I huddled shivering against a wall of rock. I drank some more of the stale, metallic-tasting water, and struggled to my feet. This time I avoided the road, skirting instead the craggy tree-hung cliff that formed the wall of the escarpment, the road running anything from fifty to two hundred feet above me. I no longer called his name—my voice was so hoarse I could hardly speak. Toward midday, as I rounded a shoulder of rock, a scrabbling rush of hooded vultures, the smallest but most obscene of African scavengers, rose up like huge vampires. A glint of sable and gold told me all I had to know. Rupert, flung from the fast-traveling van, must have gone straight over the cliff and died instantly; his neck was broken. I buried him beneath a pile of fallen boulders. Then I wandered off into the bush. I found a baobab tree, which gave me some sort of shade, and flung myself on the ground, my rifle propped against the trunk. I must have passed out for it was dark before I regained my full senses. Somehow, by grabbing at bits of vegetation and by using my rifle stock as a staff, I fought my way up the side of the ravine and caring little what happened to me or where I was going blundered blindly along the road.

I hadn't gone more than a mile or so when a cheerful, noisy truckload of Sikhs stopped and offered me a lift. To avoid hurting their feelings I climbed aboard. At least I wouldn't have to explain anything to them and they, seeing my rifle, assumed no doubt that I had been on an unsuccessful shooting expedition. At the Indian *duka*, or small general store, in

A Double Loss

Oldeani, I got a lift from Johnny Hunter, another of our rather scattered group of neighboring farmers. Johnny took me all the way to Kit'Ndovu where the Kiellands, now more or less recovered, said little but did much to cheer me up. Now, they decided, was the time to attack the crater. In a week or so the coffee would be picked and dried, and there would be a pause in the everlasting activities of the farm.

First light of the day we had chosen to begin our adventure saw Jan and myself, with rifles slung over our shoulders and butterfly nets at the ready, each of us wearing a dramatic wide-brimmed bush hat and backed up, in the true "white man in the tropics" style, by a party of African bearers loaded with cooking equipment and blankets. In single file, we moved silently off into the dappled, wild beast-haunted depths of the forest. I wondered what Gordon-Cumming, the redoubtable big game hunter and explorer, would have made of us! He would no doubt have chuckled to see us leaping wildly about, our nets flashing as we hurtled through the undergrowth in grim pursuit of some entomological wonder, caring little for the recumbent rhino that might well have been dozing behind every bush. We pushed on fast, following game trails that conveniently led ever onward and upward. The only disadvantage of these trails, which might well have been constructed by gangs of skillful engineers, was that there was always the very real possibility of rounding a corner and blundering into the rightful owners; and indeed several times we had to leap behind trees or up banks to avoid the indignant charge of a wrathful rhino. And once we found ourselves completely surrounded by a herd of buffalo, which erupted all round us and gave us an unpleasant few seconds before they went pounding off with all the noise and confusion of a defeated cavalry charge.

After hours of climbing and struggling upward we finally burst out of the thick intertwining mass of rain forest and entered the bamboo zone. Here bamboos of all sizes and heights stood like rigid ranks of lances, mile upon mile of them. Walking was easier, but there was a deep, sinister silence that was rather unnerving; no bird song, no cheerful chatter from the troops of blue monkeys which we had left behind us in the forest below. Not even the savage bark of a baboon broke the stillness. We felt, as indeed we were, unwelcome intruders in an unfriendly world. It was growing steadily colder and even the African bearers, affected by the atmosphere, stopped their eternal tuneless singing. With increased speed now we strode on upward toward the invisible rim of the crater.

Then, with little warning, we entered once more a new world. The bamboo army fell away behind us and we were in a world of sun, of heather, of weird lobelias and huge groundsel. Here, even at this height, flitted the bottle-green malachite sunbirds, and our nets were soon busy among a horde of butterflies. There were many members of the "white" family, a strange tropical variety of the clouded yellow, as well as many species I had never encountered before. We spent a long time making paper envelopes to hold our catch before starting our descent into the depths of the crater floor far below. It was a relief to be going down, at least at first, but soon the going got harder. We were coming down in a series of sliding, scrambling runs; Jan was out of sight on my right, and I had reached a particularly trying bit. The whole hillside consisted of loose flakes of lava on which it was virtually impossible to gain a decent foothold.

Idiot that I was, without a thought I grabbed my rifle by the muzzle and began using it as a support to help in my precarious downward progress. There came an ear-shattering report and a searing white hot pain shot through my left buttock, leaving me stunned and sick with fright. As I had shot myself with the sort of bullet that was designed for stopping rhino in their tracks I thought I must at the very least be in a mess, and I could feel the blood welling out and soaking my khaki shorts. The shot had brought Jan scrambling to my side and when I told him what had happened he too turned white with horror. I had been chewing a piece of gum before the shot was fired and I continued to chew on philosophically while waiting to meet my maker. Jan was the first to do something. Gingerly he peeled off my shorts, prepared for the gaping ruin he expected to find; I gritted my teeth and prepared for his verdict.

But Jan, to my amazement and indignation, was laughing almost hysterically. It seemed that the bullet had passed within a few inches of my posterior, had embedded itself in the lava and in so doing had broken off a flying fragment that had pierced my unhappy buttock, causing a bloody but entirely superficial wound. This Jan soon filled with permanganate of potash, and after a rest I was able to continue my slow, steady but damnably painful descent to the crater floor. It must have been one of the narrowest escapes in history; I might add that I no longer use rifles as walking sticks.

Flowing down the center of the crater floor was an ice-cold mountain stream, which came from a subterranean source deep in a series of grottoes leading far into the mountain side on the extreme western edge of the crater. Here in this stream I bathed my wound, washed my shorts, and was much relieved to see how slight was the damage. Truly do the gods take pity on fools and drunkards. We sat huddled in blankets close to the stream and, after a spartan meal of *posho* porridge and strong tea, I slept well if fitfully, awaking to hear the song of the night wind in the surrounding peaks and the calls of the tree hyrax in the far-distant forest. Next morning I was too stiff and sore to walk and so remained near our camp while Jan went out entomologizing. After an hour or so he returned, his collecting tin containing a number of new specimens, all of which we designated *craterensis* for want of a better appellation.

I spent the whole of that first day recuperating and listening to the sounds of the crater, which were in fact remarkably few. From the surrounding rock face came the falsetto croaking of a pair of white-necked ravens and, close to these, the squealing and grunting cries of Egyptian vultures which appeared to be nesting in the sheer precipice above us. These vultures, with their snow-white black-tipped primaries and wedge-shaped tails were impressive enough as they soared and spiraled high above us, but seen at close quarters, with their mean pinched bare faces and weak bills, they were far less attractive. Their shifty look and loose ruffled plumage somehow bore evidence to their disgusting habits and loathsome diet.

After a full day's rest I felt much more active, and Jan and I took our nets and sallied forth in search of specimens. Entering an amphitheater where the short rich grass and eternal water supply gave splendid grazing for large game, we came upon a cow rhino lying on her side while her small calf, no larger than a half-grown piglet, gamboled puppy-like around her. Squealing happily he would go bucking all over her, even climbing up onto her huge gray comatose flank and butting her with the hard round lump at the end of his nose where his front horn was already beginning to sprout. It was an enchanting sight, but we soon backed away and left them there together to enjoy the peace and quiet they deserved.

We explored the subterranean stream, following it deep into the mountain side. As we moved further from the entrance until the only light came from our small but powerful pocket torches, we noticed that a number of shallow pools had been formed in backwaters away from the turbulence of the main

A Double Loss

stream. In one of these pools we found a number of African clawed frogs, which had apparently been living in total darkness so long that they had lost all color. Even their eyes had become almost invisible; yet they seemed to be thriving and in perfect condition. Stranger still, the water was teeming with what appeared to be freshwater shrimps and other minute crustaceans, all of which were equally bereft of pigment. High overhead were colonies of bats, which appeared from below to be almost identical with the European lesser horseshoe bat; however we failed to obtain a single specimen so their identity will, alas, remain undetermined until another naturalist finds his way to this remote spot. On about our fourth day the food ran out and at 7:30 the following morning we began our long trek homeward, climbing out of the crater on the opposite side to that by which we had entered. We swung eastward and started to march. We marched all day and all night, without food, with little water, and with only a few short periods of rest, reaching Kit'Ndovu at 4:30 A.M. after twenty-one hours of practically nonstop walking. We were dead beat, hungry, thirsty but contented; our collecting boxes were full and we had defeated the Oldeani crater.

When Jan's brother Carl and his younger sister Gerd came on holiday from Norway we all went on another marathon, this time down the escarpment and out across the flat badlands toward the distant M'bulu hills. On this trip we met some members of an almost extinct aboriginal tribe that must be an offshoot of the Kalahari bushmen and who speak with the same peculiar clicking speech. These were the strange nomadic Wat'Ndiga. We were received with due ceremony by the tribal chief, who was sitting on a buckskin stool before his temporary palace (a grass hut about three feet high). His spear was stuck into the ground in front of him, he was wearing a loin cloth and very little else, and he received us with splendid pomp and dignity. He uttered the only words of Swahili he appeared to know, *Eko tombacco?* (Have you any tobacco?), and that was the sum total of our conversation. I enjoyed this safari, but I must confess it nearly killed me though it appeared to have little effect on any of the Kiellands.

Not long after this Margit told me there was a delegation of Africans waiting to see the *Bwana N'Dege* (Bird Man). The leader and most vociferous of this group had something tied up in a colored cloth. On my instructions the cloth was untied and whatever it contained deposited on the grassy bank for my inspection. This turned out to be one small and very disgruntled nestling—a lovely mottled mixture of grayish-brown and white above and narrow grayish bars beneath. I must confess I was completely foxed as to what species the bird belonged. I gave the fellow half-a-crown with which he seemed delighted, and followed this up by cursing him as a nest-robbing swine and informed him that if he brought any more such captives I would personally destroy him. He roared with merriment at the *bwana*'s waggish ways and went off happily to turn his modest fortune into native beer with as little delay as possible.

Left alone with this pathetic fledgling (now my one and only pet) I examined it closely. It also looked at me and seemed none too pleased at what it saw. I fetched some bread, milk and raw egg, a hitherto never-failing pick-me-up for almost all types of nestlings other than young birds of prey. The youngster had a powerful almost corvine beak with a slight but noticeable hook at the end that should have been sufficient clue to his identity. I opened his beak with some difficulty and pushed in some of the mixture. He looked at me in a surly way, shook his head violently and spat it out, most of it going into my face. I tried again with the same result, except that this time it all went into my face. No future here I thought. I put him into my usual small-bird-rearing compartment, the paper-lined, upturned interior of my bush hat, and went away to ponder.

An hour or so later when I looked into the hat I had a surprise. There, neat and perfect on the white paper, lay a small oval pellet, about the size of the top of my little finger. I seized the pellet with much the same enthusiasm as a prospector grabbing his first gold nugget and took it outside to dissect it. It consisted entirely of undigested portions of locusts or grasshoppers. How could I have been so ignorant? The bird before me was of course a shrike. What sort of shrike was immaterial at the moment; what was important was that he should be fed immediately. After a few failures I caught a number of smallish grasshoppers, decapitated them and removed their wingcases. I took the budding butcher bird in my right hand, and after some further manipulation managed to insert a grasshopper between the powerful black mandibles. I gave a gentle push and the insect was gone. Another followed and then, quick to get the hang of things, he began to gape and flutter his wings in the best fledgling style. Soon all the grasshoppers had gone, and the shrike seemed for the moment at least replete.

Then I had a stroke of near genius. On the terrace below reclined the *toto* (small boy) in charge of his father's mixed flock of goats and sheep. He had nothing to do all day except to see that his charges remained uneaten by leopards. I seized a pickle jar that happened to be standing conveniently on a shelf in the kitchen and I presented it to the *toto*, an intelligent lad of about twelve, indicating by word and gesture that, in return for the jar filled to the brim with grasshoppers, he would receive a modicum of largesse, and that if he didn't bring the jar overflowing with insect life he too would be destroyed. He took the jar in one dusky paw and departed for his flock, grinning like a toothpaste advertisement. He it seemed also knew the *bwana*'s little ways. That evening he returned, the bottle overflowing with assorted orthoptera. I gave him a shilling and the empty bottle. He was happy, I was happy, and above all so was the shrike, who promptly ate himself into a blissful state of lethargy.

Resorting once more to Roberts' *Birds of South Africa* I quickly discovered that my new protégé was a young fiscal shrike, popularly known in the Union as a bullhead, butcher bird, or Jackie hangman, all descriptive but hardly flattering designations. I called him Dingaan, after a singularly bloodly-minded Zulu chief, and hoped he wouldn't live up to the name. Dingaan soon became a great and highly appreciated addition to the sadly depleted household. What he lacked in size he more than made up for in sheer effrontery. After the first few days he became utterly fearless. He would sit on my fist like a miniature hawk, or on my shoulder like a tame jackdaw, and he would follow me everywhere. He was in fact an endearing and highly individual mixture of kestrel and crow, with some outstanding characteristics purely his own.

The adult fiscal is probably one of the most outstanding and ubiquitous bird personalities over the entire southern half of Africa. It is also a particularly handsome one, with its glossy black and white plumage, and its long magpie-like tail. It is a small bird, only about eight inches long, but suffers from no inferiority complex, perching conspicuously and piratically on every telegraph wire and acacia bush, whence it keeps its far-seeing black eyes wide open for any large insect, lizard or small rodent that is unwise enough to leave cover. Dingaan's juvenile plumage was rather like that of a very pale hen pheasant, all wavy lines, blotches and bars. From the day he could fly he was allowed complete liberty and when following me would flit restlessly from tree to tree calling in his strident unmusical shrike voice. His appetite was phenomenal; he could stow away at least eight huge locusts and in an hour or less

A Double Loss

would disgorge a hawklike pellet and start stoking up again. The *toto* with the pickle jar was kept hard at it and he certainly earned his shillings.

Dingaan was to prove one of the most intelligent and lovable of all my bird acquaintances, and he very probably saved my life, possibly twice. During this period I spent an increasing amount of time on the estate which I hoped might eventually be mine. One morning I started off armed only with a shovel and a butterfly net and with Dingaan as usual in noisy attendance. I carried the shovel to repair the furrow that had been trampled and generally mucked up by a herd of thirsty elephants. I worked hard and soon had repaired the damage, while Dingaan amused himself with catching insects or disporting himself at the shallow edge of the furrow, for like most birds he dearly loved a bath. The job completed, I sat on the grass verge to enjoy a cigarette. Idly watching the butterflies playing about the flowery bank opposite I spotted a species of swallowtail I had never seen before and, grabbing my net, was on my feet in a trice. The swallowtail was dancing about a gap in the forest wall a dozen paces ahead of me as I stalked forward with poised net.

Dingaan was sitting on a tendril of tropical convolvulus watching me and at the same time alert for any entertainment, gastronomic or otherwise, that might be going. Suddenly he shot forward like a stooping kestrel, and began to shriek in a voice totally unlike his usual querulous complaint. Diving and swearing at something unseen by me in the gap ahead, he was obviously both enraged and panic-stricken, becoming more and more hysterical. I stopped dead in my tracks and peered forward; only feet away and level with my own I saw a pair of jade-hard unwinking eyes, set in an evil diamond-shaped head, as dark or darker than the shadows from which it arose. It was the head of a black mamba, Africa's most deadly and most aggressive snake. Seldom have I felt so frightened and so helpless. Keeping my head and torso as still as possible I moved backward a slow foot at a time, until I felt the distance between us had widened enough. Then I turned and did an Olympic type jump for my shovel, which luckily was close at hand. Thus armed my confidence returned; I picked up a huge slab of rock and hurled it at the gap where the mamba still lurked. There was a crash followed by a dry slithering sound as the snake dropped from the bushes and made off into the recesses of the forest from which it had come. I lay on the grass and was nearly physically sick with fear and relief. Dingaan, convinced he had routed the evil monster, gave up swearing and hopped onto my hat, his late nursery, for a snooze.

There must indeed have been a curse on that irrigation furrow. A few weeks later, the snake incident no longer prominent in my mind, I started off again to examine the life blood of what I was still determined would be my home and livelihood. I had slept badly the night before, having had an evil dream about being charged by a rhino that I remember to this day. However, I set off with my net and shovel as before, with Dingaan riding happily on my shoulder or flickering from thicket to thicket ahead of me. I seldom carried a rifle; I was not an expert shot, and I had a misplaced theory that unwounded game seldom attacked unless provoked. Once more I cleared the furrow at almost the identical place where the snake had been. The sun was just beginning to sink below the crater rim as I stopped for a rest and a well-earned smoke.

Dingaan was on his tendril look-out post, holding a dung beetle in his feet and dismembering it with his beak in truly hawk-like fashion. I stood up to go and as I did so Dingaan started to shriek and swear at something immediately behind me. I glanced back and instantly hurled myself sideways as the

buffalo charged at close quarters, hitting me in the shoulder and hurling me face down on the ground. Had it not been for Dingaan I would in all probability have been hit in the back and that would have been that. Instinctively I turned on my back as the buffalo spun round and came in again. Somehow I was able to bring my feet into play against the great boss of his horns as he pressed me against the bank, so that he was unable to get them under me to throw me into the air. My boots slipped off the boss and he drove at my left leg, impaling it. I grabbed his horns and kicked him in the face with all my power. I could feel the skin of my palms tearing on the emery-paper-like surface of his horns as he worked one like a corkscrew into my leg.

He withdrew the horn, ripping a hole in my flesh, and stood back panting. He was too close to get up much speed as he came in again and I was able somehow, though exactly how I shall never know, to dodge his pounding hooves and hurl myself behind him. Before he could turn again I threw myself into the thicket, ripping my clothes to shreds. I saw a branch, grabbed it, and scrambled up as he swung round and came back, looking for me to finish me off. I was only a few feet above his back but that was high enough. He stood under the tree looking up at me and my nostrils were filled with the sickening odor of the beast. Suddenly he swung round and went plunging off into the forest and I could hear the crash and rustle of his departure for a long way as he retreated.

I lay for a long time astride my branch chilled and feeling one grinding aching lump of pain. Semiconscious, I gripped the bough on which I lay until the shrill pleading call of Dingaan, somewhere unseen in the darkening forest, recalled me to full awareness. I answered and at length he came to me and nestled close beside my left cheek. Somehow I managed to lower myself to the ground and set off homeward, half limping, half crawling, using a branch I had picked up from the ground to support myself. It was a nightmare journey. Every rustle in the bush could have been that accursed buffalo coming back to finish his work. All the pain-wracked aching way to the farm road Dingaan perched on my shoulder, his body warm and comforting against my cheek. At last we reached the road and I sat by the side utterly done for. Mercifully I soon heard the rattle and roar of an approaching motor and saw the welcome glow of headlights. Within twenty minutes I was tucked up in bed in the hospital at Oldeani with Dingaan, like Cressida before him, perched on the bed rail. Under the care of the Polish doctor and the nurse I was soon as fit as ever, but the scar on my left leg is a constant reminder of what was undoubtedly the closest shave of my life. Few have wrestled single-handed with a buffalo and survived.

7

Audax Joins Dingaan

A WEEK OR SO after my return from the hospital I received a letter from the principal of the agricultural school requesting me to attend the beginning of term in three weeks' time. The next two weeks passed quickly. I was in many ways sorry to leave the Kiellands and their home, but I was looking forward to doing something positive and, besides, I had been asked to return to Kit'Ndovu for the holidays. I had made up my mind to take my time over the journey to Nairobi, and thence to Njoro where the college was situated. A week before I was supposed to report to the school Jan drove me to Oldeani, where I caught the usual disreputable Indian bus to Arusha. My entire collection of worldly goods was crammed into one smallish case, which had the embarrassing habit of swinging open at the most awkward times and decanting my few somewhat shabby possessions into full public view. Whereas for some years I had been traveling with a dog and two fairly large birds I now only had Dingaan, who could if necessary snuggle, tail and all, quite comfortably in the pocket of my bush jacket.

Arriving in Njoro, a few miles from Nakuru, the evening before the term was due to start, I was shown to my humble but

Audax Joins Dingaan

adequate bachelor quarters (one small room in a row of identical ones), unpacked my luggage, which included a jar of locusts, settled Dingaan on a suitable perch and prepared to see what fate had in store for me. The Egerton School of Agriculture, named after Lord Egerton of Tatton, a well-known early settler and one of the largest landowners in that part of the world, had not been long established when Dingaan and I arrived. It was a pleasantly laid out, efficiently run establishment, and it only took a few hours before I realized that I was going to enjoy myself immensely; and what was more I might even learn something. The students ranged in age and experience from a somewhat dyspeptic ex-wartime brigadier in his fifties to the sons and daughters of local settlers still in their teens; from the earnest and dedicated to the frivolous and unashamedly bone idle. I was due to take a year's course at the end of which, if I had worked hard enough and passed enough exams, I would become the proud possessor of an agricultural certificate which, abbreviated to Cert. Ag. Ege., would be, I was assured, a passport to many interesting and remunerative jobs.

My fellow students were a pleasant and affable bunch. Oddly enough, despite my advanced age (I was just over thirty) I seemed to gravitate socially to and be accepted by the younger set rather than by my contemporaries, who were among the more serious-minded and conscientious of my classmates. Besides, these all seemed immensely rich and apparently owners to the last man of huge estates of their own. One lad, Michael Hogg, who hailed from Eastbourne, owned a splendid Model A Ford, which apparently ran entirely on lamp oil. He and his family had a farm near Thomson's Falls, and the first weekend he invited me and a few others over to his place which, after a somewhat hazardous journey in his courageous but dilapidated vehicle, we eventually reached.

Thomson's Falls was the center of the White Highlands, the

best farming land in Kenya. It stood about eight thousand feet above sea level, and was so English-looking as hardly to be true. All that belied the British atmosphere was that in tall creeper-festooned woods, nestling between valleys as red as those in Devon or South Wales, lived troops of the beautiful black and white long-furred Colobus monkeys, handsomest and most gentle of all the primates. Over the great wide expanse of grazing land, where were pastured herds of imported white-faced Herefords and shorthorn cattle, there hunted the splendid black-and-white augur buzzard, perhaps the most characteristic of all up-country birds, whose clear challenging call, AUNG, AUNG, AUNG, so unlike the British buzzard's peevish mew, makes me think nostalgically of those clear-aired sweeping spaces to this day. Dingaan took happily to school life and not even Mrs. Landevaal, the rather austere German housekeeper, objected to the presence of this unorthodox guest.

In a remarkably short time he had developed a marked personality. While I studied he would wander the grounds and he spent much time perched aloft on the telephone wires, where he was soon challenged by, fought with, and as often as not defeated the local fiscals, who resented the presence of this wanderer from the far south. A number of my fellow students shortly afterward appeared with nestling shrikes, allegedly found abandoned, but in reality I suspect taken from nests of which there were a number in the vicinity. Thus was formed the Egerton Shrike Club; later, when an augur buzzard and a steppe buzzard were added to the collection, to become the Egerton Falconry Club, probably the only falconry club in the whole of East Africa at that time.

I thoroughly enjoyed my year at the school, and I made some good friends among the locals I met. I especially recall the Pearce family. Tam and Jeanne were real settlers of the old school, who farmed one of the best farms in Kenya. Their daughter, universally known as Blue for some obscure reason,

Audax Joins Dingaan

worked in what would now be known as a boutique in Nakuru. She later married Brian Pockley, a neighbor who owned the most magnificent Arab stallion I have ever seen. I spent several weekends with the Pearce family and great fun they were too. Another almost legendary family was the F.O'B. Wilsons, who had an immense estate, some twenty or thirty miles from Nairobi, off the Mombasa road. Their way of life was uncannily reminiscent of that of the old-time Western cattle barons, even to their virtually living in the saddle and carrying six-shooters (as a protection against the Mau Mau, who were at that time just beginning to create a nuisance).

After a few weeks at the school I received an eyass augur buzzard in a very downy stage, with only a few dark feathers protruding from the cotton wool on its back. It speaks much for the tolerance of the staff and my fellow students that no one seemed to object to my turning my room into a cross between a mews and an aviary. Anyway, Audax soon proved worthy of his big reception. He was big, fat and helpless, and as appealing as are all young birds of prey. I was delighted to see him, but Dingaan made it blatantly obvious that he was not.

Audax did well on a diet consisting almost exclusively of mole rats, grew apace and was soon as ridiculously confiding as are nearly all hand-reared hawks. He was flown "at hack" about the grounds, was quickly trained to the lure and fist, and would fly very prettily indeed, showing a good deal more drive and punch than is usual among European buzzards. In his new coat of feathers, slate-gray on the back, his breast white with just a touch of burnt toast, and a rich sienna tail, he was indeed a handsome beast. He and Dingaan accompanied me when, the first term behind me, I returned to KitN'dovu, to the Kiellands and to my butterfly collecting. I am afraid that Audax wore his welcome a bit thin by polishing off an entire brood of the undersized bantam-like fowls that belonged to the *neopara*, or native supervisor, at the farm, but at least this showed how predatory an augur buzzard can become.

The high spot of this particular holiday was attending an Afrikaans wedding ceremony of Piet van de Merwe to Catrina du Toit. We put on freshly laundered bush jackets and khaki slacks, the only clothes any of us seemed to possess. The Kielland girls however, in all their feminine finery, were quite an eyeful, as hitherto I had only seen them dressed much the same as ourselves. Fun was fast and furious; the party started at about 4:00 P.M. and was still going strong at 10:00 the following morning, by which time I for one had fallen by the wayside. What happened to the bride and bridegroom was anybody's guess, for I cannot recall having seen them at all. I paid no further visits to the farm that I had so longed to own, as after the buffalo episode I concluded that something or someone around there didn't care for me, and in any case the urge to farm my own land had weakened as the months passed. I was really pleased to return to the Egerton School, and to meet my comparatively new friends again.

During the second term the great agricultural show took place in Nakuru. It was one of the chief social and agricultural events of the East African year, and all the students of the Njoro School were dragooned more or less willingly into helping. I got conned into stewarding in the show-jumping ring, which seemed to me at least a cut above the more mundane sheep and cattle exhibits. I am unlikely to forget that show. Dingaan, together with a dish of grasshoppers and other delicacies, had been shut in my room for the day. Audax the augur, however, had been absent when we left to go to the Nakuru show ground a few miles away. Since he used to range widely over the surrounding country, I gave him barely a thought. The morning passed uneventfully; there was plenty to do and to see.

Audax Joins Dingaan

At about three o'clock I was standing in a corner of the main ring, my duties temporarily discharged, when high in the sky I heard a faint AUNG, AUNG, AUNG; glancing idly skyward I saw far above me the silvery breast of an augur buzzard. They were fairly plentiful thereabouts, so I paid scant attention until a minute or so later I heard it again, much nearer and clearer. I looked up; cruising just above the grandstand, his jesses streaming behind him, floated Audax. With hardly a thought I removed the red silk scarf I was wearing and knotted it into a ball. Seizing a piece of twine I attached it to the scarf, swung this makeshift lure above my head, and shouted. Audax keeled over at once, half closed his wings and plummeted into the arena like a cannon ball. I kept the lure just out of reach as he climbed like a falcon to turn over and stoop again and again. Finally I threw the lure high in the air, where Audax caught it, brought it to earth, and stood over it mantling magnificently. The audience was staggered by this entirely impromptu display, and I was told by many that for them it was the high point of the whole show. The principal of the school was understandably less amused, but took it very well indeed. After this I put Audax safely in the school truck and hurriedly sent one of the African staff, who happened to be standing by, into Nakuru for a piece of meat as a well-merited award.

One weekend a group of friends and I climbed to the rim of Longanot crater to look for lammergeiers. We didn't see any, but I did collect a pair of young triangular-spotted pigeons from the loft of a house at Gilgil. These pigeons, which are like exceedingly glamourized red chequer homers, with bare red rings around their eyes giving them a peculiarly belligerent appearance, are both common and in fact almost semidomesticated in many parts of Africa. I sometimes wish I could have brought a few pairs back to Britain with me, as they would assuredly add tone to any pigeon loft. These pigeons were kept in a rapidly constructed shed, which served well enough as a temporary dovecote; when released they stayed about for some time before gradually drifting back to the wild from which they had come.

Dingaan meanwhile had reached full maturity, moulted into his dramatic pied plumage and, among other accomplishments, had become a singer of no mean talent. In fact for tone, variety and endurance he could have put many a vaunted British songster on his mettle. He also entertained me hugely when I was writing up my notes in the evening and hordes of sausage flies would enter the open window and blunder brainlessly about. Dingaan, perched on a lintel or window ledge, would dart at and seize in full flight as many of these unattractive beasts as his considerable appetite could manage. Sometimes a fast-flying noctuid moth would appear, and a sort of aerial rat hunt, under and around the furniture, would ensue. This exhibition of indoor falconry in miniature proved a highly popular entertainment, and my room was often crammed with spectators who called in to see Dingaan do his stuff.

So the year passed swiftly, punctuated with a few memorable incidents, such as when I caught an immense python that was bathing in the cattle trough on the farm of the Cook family just outside Nakuru. This ferocious-seeming monster proved surprisingly easy to subdue and was placed in a sack before being ultimately released elsewhere. In the course of the brief struggle the brute bit me savagely in the hand. The wound later turned septic, owing no doubt to the foul condition of the beast's teeth, but to my intense relief he made little attempt to engulf and crush me in his coils. My stock went up at least among the few African onlookers, who could not be induced to approach the monstrous *nyoka*, even when it was safely incarcerated in the sack.

Audax Joins Dingaan

The course came to an end far too quickly. I remember my last evening at Njoro, sitting on an outcrop of rock high above the school farm while Dingaan grubbed about for insects in the rough herbage and Audax sat sentinel-like in an acacia tree above me. I looked out onto what I still believe to be one of the loveliest landscapes in the world. Behind me rose the dense cedar forests of Mau summit; below me stretched a great hazy blue-gray bowl of rich farmland and acacia-dotted bush over which, twinkling in the evening light, hovered and glided a pair of the graceful black-shouldered kites. The sweet fertile undulating countryside rolled away before me until it climbed to meet the sky away to the southeast, where rose the gentle grassy slopes of Rongai crater, the living heart of Kenya as I knew it then.

The next day found me in Nairobi at the home of my friend Bill Harrison and his family. We were jobless and we had heard that the East African High Commission needed field volunteers to control the swarms of desert locusts that periodically ravaged the countryside. Therefore, a few days after leaving Njoro we presented ourselves before the Chief Locust Officer, an affable gentleman named Bernie Fickling. After a few questions and telephone calls—made no doubt to check our credentials—we found ourselves accepted, and with remarkably little delay we had been kitted out with tent, camping equipment and best of all a pay advance. We were soon on our way in a land-rover to attend a conference at Isiolo, at the edge of the northern frontier division of Kenya.

Bill drove and I sat beside him with Dingaan on my knee and Audax perched morosely on a crate in the back. Also crouching in the back was Zacharia, until recently faithful Kikuyu houseboy to Bill's mother, and now on his way with us to act as cook, so-called mechanic and general factotum. I shall long remember that drive. We were free, young, we had money in our pockets and we were heading for the loveliest and most exciting country in the entire world—or so we thought at that time. The land-rover climbed, descended and climbed again; the great icy miter-shaped peak of Mount Kenya rose before us. We reached Nanyuki, where we spent the night before beginning the long run down to the arid, almost waterless semidesert below; a wilderness stretching mile upon blistering mile right up to Somalia hundreds of miles to the north. It was a country I had always had an urge to visit and now I was on my way there as guest of the Kenya government.

It seemed that the periodic conferences of the Locust Control, in which that organization's top brass met the lower echelons, always took place at similarly outlandish venues. It was a weird experience. Finding our way to the place where the conference was due to be held, we parked ourselves under a suitably shady acacia tree, put up our tent, drank a few bottles of Coca-Cola (that eternal standby of the white man in the tropics) and took stock of the situation. The locust officers were gathering like vultures to the kill, from Lodwar, from Moyale, and from Marsibit, from Garissa in the east and from Kitale in the west. Of necessity they were a wild bunch of individualists to cope with such a job. Loneliness and hardship would be their companions, month after month, in the heat of the African sun and the bitter cold of the desert night, with only natives to speak to. As I looked around me I wondered what inner resources would these men need to keep them going? Later I came to know some of my fellow locust-hunters well and found them, despite their wild appearance and extraordinary dress, to be likable, well-balanced humorous fellows, neither antisocial nor particularly eccentric. Bernie Fickling, the great man himself, arrived—the typical higher government servant on safari, immaculate in khaki and snow-

Audax Joins Dingaan

white topi. And so the conference opened. It was all rather a lark; none of those present was likely to be overawed by authority and it was, I remember, as much a social gathering as anything else. We listened to talks about the various "instars" or stages of locust development during the morning and we lounged and bathed in the unexpectedly cool and refreshing pool in the afternoon.

A common bird there was the small rusty black fantailed raven, so weedy and furtive compared to the spectacularly powerful and arrogant British raven, the undisputed master of our Welsh hillsides. These so-called ravens, actually not much bigger than the more abundant pied crows, were very tame and would scrounge scraps left outside for them in a rather apologetic, almost pitiful, way. They seemed to suffer from the heat and were almost continually gaping and panting with wide-open beaks, a peculiar characteristic in a largely desert-haunting species. The fiscal shrikes had disappeared, their place being taken by the much larger, handsomer and more sociable gray-backed fiscals, which gathered in noisy flocks on any prominent bush. These would fly off on being approached with a rather weak, drifting flight in which their disproportionately long tails seemed to take over, the slightest gust of wind blowing them off course. This worried them but little; they never seemed unduly anxious to go anywhere in particular. They reminded me of a scattered flock of fieldfares trailing indecisively from one hawthorn bush to the next.

The conference over, we returned to Nairobi where I found to my delighted amazement that my mother had arrived the night before and was staying at a hotel in the city. I was overjoyed to see her again after nearly five years. The next morning I reported to the Desert Locust Control office just behind the Coryndon Museum, where I was supplied with a land-rover and told to make my way to Kajiao, an outpost in the Masai district about half-way between Nairobi and the Tanganyika border. I now had a land-rover and an ignition key but there was a major snag: I didn't drive! I had never had the slightest wish to drive a car, and up to that moment I had not been unduly inconvenienced. Once again fortune came to my aid. Bill Harrison very kindly allowed me to take on the services of Zacharia, who *could* drive. And I was determined at last to learn how, necessity being a powerful persuader.

My mother naturally wanted to come with me. So I called for her at her hotel and soon we were all hammering our way over the dust and corrugations to Kajiado, where I chose the most attractive spot I could find and made camp. The next day, with the long-suffering Zacharia as instructor, I took the land-rover out into the middle of the Arthi Plains and learned to drive. Hour after hour I practiced and was soon confident enough to try her on the open road. Within a week I felt as if I had driven all my life; the fact that I had passed no test and held no driving license worried me not one whit.

It was time, I decided, that I began to earn my living and to find some locusts. With my mother as navigator and Zacharia, Audax and Dingaan as crew, we loaded up the land-rover and set off on safari. Anything that becomes part of my life, whether it be bird, beast or automobile, must have a name, and so KBE25 became, for the next few months at least, Laura the land-rover. The flag she flew from her hood displayed the figure of a locust, rampant of course, carefully and accurately fashioned from the finest silk the local *duka* could supply. Thus adorned, we set forth for the furthest boundaries of my new territory, picking up en route at the Mananga River Hotel one Fergus McBain, a cheery Scotsman, lately employed as a sort of driver-cum-game-spotter by Budge Gethin at Rhino Camp. I had known Fergus for some time and had much in common with him, greatly admiring his ability to

Audax Joins Dingaan

quote faultless pages from Henry Williamson's *Dandelion Days*, which he indulged in at every opportunity with the slightest encouragement. He was a splendid companion and a lover of all beasts and birds—wild, tame or domesticated. He also, like me, had a decided weakness for the fleshpots and was utterly at ease and relaxed in any sort of company. Fergus was anxious to join the Locust Control and so I agreed that when I next called in at the office in Nairobi I would take him along and introduce him.

Meanwhile here we were a merry bunch of adventurers, rocking, sliding and jolting our way across the Amboseli Reserve on our way to Laitokitok. My mother was well over sixty but she enjoyed the excitement and variety of the trip as much as anyone, and put up with the inevitable discomfort—heat, dust and flies—as cheerfully as she must have done over thirty years before when she and my father had been among the few white people in Somaliland. She had never been in East Africa before, except for a brief sojourn in Nyasaland, and was thrilled with the wildebeest, beisa oryx, roan antelope, lesser kudu and the many other exciting and interesting animals she saw. When an unfriendly rhino charged the land-rover, nearly overturning it, she was more delighted than dismayed. Nature put on a gala performance for her and she saw a rare sight: a leopard stalking, springing upon and carrying off its prey, a Grant's gazelle fawn. She saw more than one pride of lion, not the lethargic semitame beasts of the Nairobi National Park but the real thing, tawny, black-maned and redolent with controlled savagery.

The first night of the trip we were in the government rest house at Laitokitok, high up beneath the brooding shadow of Kilimanjaro. We awoke to the wheezing, drawling voices of the go-away birds, a kind of plantain eater or touraco, which has the unpopular habit with hunters of appearing from nowhere just before a splendid shot is offered and shouting out in their infuriating affected voices, GO WHEE! GO WHEE!, which invariably puts the hitherto unsuspecting victim on the *qui vive*. We were lucky because, although we saw no locusts, we did see a striped hyena, a smaller, rarer, and rather more attractive animal than its spotted namesake. We also saw a sow warthog with a litter of tiny piglets which, catching sight of the advancing land-rover, stuck their ridiculous tails in the air and belted across the open salt-laden bed of a dried-up lake until they saw a convenient antbear earth. Into this they plunged backward, scuttling into the welcoming darkness tail first, the old sow going in last, her unprepossessing face with its vicious tusks forming a barricade few predators would tackle lightly.

I wired a perch for Audax on the hood of the land-rover and after a good deal of trial and error managed to persuade him to hold on while I drove it slowly down the Arusha-Nairobi road. I had seen a flock of guinea-fowl or *kanga* gritting there one evening and wanted to try an experiment. I had shot a few with the new twelve bore that my mother had brought me from home, and had given Audax one when he was hungry. He made short work of it. I made a lure of guinea-fowl wings and fed him with bits of meat fastened thereon. Now I was ready for the trial. I drove slowly and carefully down the road in the golden light of early evening. Rounding a corner, we saw a flock of guinea-fowl squatting and shuffling in the thick white dust. In bottom gear I slid forward; the flock was about thirty or forty yards away when Audax cottoned on. Swiftly and silently he launched himself from his perch. The guinea-fowl took to their legs as Audax hurtled into their midst and cries like indignant machine-guns exploded into the air while Audax landed on the road verge looking puzzled and hard done by. The next day I kept him sharp set and tried again. He left the perch quicker and, traveling faster, he followed a youngish *kanga* into the bush. After a terrific commotion I found he had pinned it to the ground. I went in and dispatched it at once. That evening I rewarded him well. Later, at long intervals, he caught three more guinea-fowl, one in flight, and a number of *kwali*, or francolin partridge. I had proved my point: an augur could, if properly entered, take game birds. In fact I would say that they are as useful as the better-known red-tailed buzzard of North America.

After my mother's departure I was consoling myself with many cups of strong coffee when I was joined, as I sat morosely at my table in the lounge of Torr's Hotel, by Diana Hartley, a member of a family consisting almost entirely of white hunters, big-game collectors and other colorful characters of a type which East Africa seems to produce in marvelous variety. I had met her once or twice before but now, it seemed, she had a problem: a three-foot-long African crocodile with jaws and teeth like a Staffordshire bull terrier or worse, and with a temper to match.

This delightful beast had been presented to Diana the night before by someone who, knowing her weakness for this sort of thing, had hoisted it out of the Tana River and brought it into Nairobi in the certainty that such a windfall would be acceptable. So it would have been normally; but as it happened Diana was off on safari somewhere to the west and had no one to act as temporary crocodile-sitter—until she saw me nursing my wretchedness. Within minutes I had been conned into taking this loathsome leviathan to my camp at Kajiado where, reclining in my canvas bathtub, it began a reign of terror that lasted for nearly a month. Seldom have I been so relieved as when Diana arrived to retrieve this unlovable saurian. "You haven't been giving him enough roughage," she remarked ungratefully, as the croc leered at her from the stagnant depths of the tub. Making a conciliatory gesture with her left hand, her newfound friend lunged at her, missing her fingertips by inches. "Over to you, chum," said I, as we tipped the horror into the back of her truck, where it reclined immobile in a bed of wet grass specially prepared for its reception.

8

Locust Hunter-at-Large

SHORTLY AFTER this episode a large number of huge Chevrolet trucks were received by the Desert Locust Control headquarters at Nairobi. These lumbering mastodons, it seemed, had been sent out to the Middle East to serve as troop carriers during the war, but for some reason were never used and had been gathering dust and rust ever since. It had been decided that they were just the thing for conveying stores, men and bait about the locust-infested areas of East Africa. As foul fate would have it I was called upon to take this untested convoy out on a trial trip to see how many would give up the ghost en route. Unfortunately I was not in sole charge of this expedition, supreme command having been given to a recently retired captain of the Royal Navy. Regrettably Jolly Jack and I didn't hit it off from the beginning; this bluff hearty seaman failed to remember he was no longer on the bridge and that I was not a cowering newly joined midshipman. The resulting considerable discord was in no way dissipated when the Chevs began breaking down, left, right and center.

These trucks were driven and maintained by a gang of bolshie Kikuyus, who were, I suspect, hovering on the brink of the Mau Mau movement. That trip can have done little to persuade them back to the path of righteousness. Unable to hang them from the yardarm Jolly Jack contented himself with cursing everyone in sight, myself included. When I told him in a few brief well-chosen phrases what I thought of him and what he could do with the convoy he, as my old platoon sergeant would have put it, "went off alarming," and threatened to have me court-martialed or the equivalent immediately we returned to Nairobi. When I failed to blanch with terror at this prospect he gave up, and for the remainder of this epic journey we hardly spoke to each other. It was not a happy trip, but it served its purpose. The trucks were weighed in the balance and found to be exceedingly wanting.

When I eventually got back to Nairobi I considered giving in my notice, but after a chat with Bernie Fickling, who it appeared had much the same views of the captain as I did, I changed my mind. I decided it was time I took a few days' leave and was about to depart when the first locust swarm appeared in my tom tom, like a million squadrons of minute hostile yellowish aircraft. They flew mindlessly, steadily past, heading southward at a steady unwavering five miles per hour. In their wake came the locust eaters, the black kites, harriers of several species, squadrons of white storks, secretary birds and maribou, while up above flickered and glided the falcons (kestrels, lesser kestrels, red-footed falcons and hobbies), all coming to feast on this entomological manna. It was an ornithologist's dream come true, but I wasn't employed to stare at birds, however richly represented. After sending a report to HQ as to the estimated size (in square miles), direction and speed of the advancing army, I had to follow, to keep in touch with and eventually to kill the hoppers as they emerged and, wingless, started marching remorselessly onward. At the back of my tent I stored the weapons used to fight this implacable invasion, sack after sack of bran mixed with Gamaxene, which was to be laid out in lines on a wide front to tempt the appetite of the uncountable millions.

A few days after my message was received in Nairobi I was relaxing after a hard, hot and thirsty morning when, with a crunch of wheels crushing the dry acacia seed and a metallic groan from ill-used gears, my opposite number from Arusha drew up beside my tent. Paddy Fitzpatrick was a type of man with whom the outposts of Empire used to abound. A lean sun-dried Irishman, he had won a particularly good MC when in Burma. Captured by the Japanese he had turned the tables, killing some of his captors with their own weapons, escaped and marched through the jungle with no food and only filthy streams from which to drink, until days or weeks later he caught up with his own HQ. "Fitzpatrick reporting for duty," he is alleged to have remarked to his astonished C.O., with an adjutant's inspection-type salute, after which he collapsed upon the floor. He had never adapted to post-war Britain and, like many of us at that time, had tried a number of jobs before drifting to East Africa and to the unconventional life of Desert Locust Control whose lack of confinement suited him well.

With Fitzpatrick was a group of temporary locust officers, somewhat inarticulate and decidedly Anglophobe Afrikaans, who showed in no uncertain manner that they did not care for the *Verdommt Roinek* and his ways. Fitzpatrick and I finished a bottle of brandy and discussed the coming slaughter while the budding Krugers outside sat on their bedding rolls and glared at us. I could see this was going to be a jolly party. I poured a cup of coffee and took it outside to the apparent leader of the Dutch faction; he made rude noises in his throat which I took to be a refusal and continued to scowl ferociously at the inoffensive skyline. After a few well-chosen remarks on his civility I returned to Paddy who, it seemed, being an Irishman had not incurred quite such a degree of dislike. From then on, other than to mime what I wanted done, I ignored them.

The tiny hoppers, black and wingless in their first nymphal state, were everywhere. In the warm hollow depressions where the eggs had hatched every patch of vegetation, every bush and low shrub, swarmed with their clinging masses. There was something so repulsive about them as to be almost frightening. Perhaps it was the immensity of this mindless multitude that so affected me. I wanted to kill, to wipe out, to utterly destroy! But we had to wait a little while until the locust hoppers had all hatched, had formed into companies, battalions and divisions and moved forward, led on by deeply implanted instinct. Then it was time to act. We slit the sacks and spread the "ammunition's" lethal contents in the path of the advancing enemy. Locusts will eat anything, even wash on the line; they ate this and perished by the millions. We must have scored at least an eighty percent success. Those that we didn't kill were cleaned up by the birds, jackals and mongooses; it was an overwhelming victory.

Fitzpatrick and his gang went back to Arusha and I returned to Kajiado. With the blessing of the High Commission I recruited a large number of part-time amateur locust spotters. The duties of these Masai herdsmen consisted of keeping a watchful eye open for signs of *uzigi* activity and reporting to me, the *bwana uzigi*, at my camp. These amateur locusteers were put on my pay roll and paid monthly wages for their services. The whole system was delightfully easy going; I collected large sums of money from HQ, and with this I paid my assistants, who duly signed for this largesse in the appropriate spaces opposite their names. The only snag was that their names, being Masai, all began with Ol, and were virtually impossible to memorize. I wrote out these names as accurately as I could and the recipient duly pressed his thumb onto an ink pad, then onto the page. Had I been a dishonest fellow I could no doubt have thought out a suitable number of imaginary Ols and, pressing my own thumb into the corresponding spaces, rewarded myself handsomely.

On one occasion I went to deal with a locust swarm in the Garissa area, miles away on the border of the northern frontier division, though I can't remember why. Crossing the bridge over the Tana River I all but got myself shot by a wandering patrol of Somali Askaris. When I complained indignantly to the D.C., one Chevenix-Trench, about this unnecessarily fervent reception, he replied that I had only myself to blame for not announcing my proposed visit beforehand. Furthermore, he expressed his displeasure that I had been rude to the sergeant in charge of the patrol, as indeed I understandably had been. When the D.C., peering into the back of my land-rover, spotted Audax perched upon my bedding roll, his attitude changed at once, although I must confess I still felt a bit huffy; after all it was I who had just been all but perforated by quantities of 303 ammo. However falconry is a great leveler, and we were soon deep in conversation about Audax's beauty and prowess and many other aspects of the noble art. Incidentally, years later, I heard that a certain Chevenix-Trench had introduced falconry into the curriculum at Bradfield, of which he was then headmaster, and I wondered if there was any connection with this rather inauspicious meeting on the banks of the Tana River.

Our first slight misunderstanding having been cleared up I had a pleasant and hospitable visit to Garissa, though whether I accomplished what I had been sent out to do I cannot for the life of me remember. I do, however, recall long friendly discussions over a much-appreciated sundowner as we sat aloft on the flat roof of the castellated Foreign-Legion style D.C.'s house, while the cicadas buzzed and the desert wind played among the borassus palm leaves below.

On the second evening, Fergus McBain appeared. He had

been paying one of his not infrequent visits to Malindi, and was feeling the worse for it. I doubt if the ensuing few hours did much to assuage his burning throat and fevered brow, but he was a resilient type and somehow managed to stagger forth at dawn the following morning to watch Audax take on the big vulturine guinea-fowl that lived in scattered coveys among the virtually impenetrable wait-a-bit thickets. They would issue forth at dawn and at dusk to drink at probably the only available pool for miles and, after drinking their fill, would drift somewhat aimlessly about looking for the insects and vegetable matter on which they fed. They were heavy birds, considerably bulkier and taller than the more common crested and helmeted guinea-fowl of the East African highlands. They were well named too; with their long cobalt-blue necks and predatory bare heads they did indeed look like glamourized vultures, and I was doubtful whether Audax would have the courage to tackle so robust and self-confident an opponent, especially as he, a native of the cool, misty, forest-crowned highlands, was showing signs of disliking the intense heat.

However, when I went to collect him, he showed some enthusiasm and began his war dance, jumping up and down on his perch and calling loudly to show that he was ready for action. I slipped his leash from the perch and in a moment he was on my fist, gripping hard enough to make me thankful that I had recently bought a new hawking glove, well-made for me by an Indian craftsman. I sat in the passenger seat of the land-rover and we raced away across the sandy corrugated wasteland, startling a small group of home-going gerenuks, the graceful chestnut-hued giraffe antelope. Their slender necks, length of leg and wonderful balance enable them to stand upright on their hindlegs to nibble delicate shoots a surprising distance above their heads. From far away came the staccato call of guinea-fowl and Audax tensed, gripping my gloved fist hard as he roused, preparing himself for the forthcoming encounter. I slipped the leash from his jesses as the land-rover bounced over the scrubland. Ahead was a wall of wait-a-bit thorn, tunneled and intersected by game trails, the secret highways of dik dik, oribi and duiker. Rounding this prickly promontory we saw, a hundred yards or so ahead, a scurrying file of guinea-fowl, their spangled plumage appearing almost black against the desert background. Audax needed no bidding but hurled himself forward and, with a couple of powerful downward strokes, carved himself through the shimmering air toward his quarry.

With a mocking falsetto croak a dark object detached itself from the lower branches of a stunted tree and dropped into his flight path. The white-necked raven had seen Audax's approach and was determined to spoil his sport; all birds of prey were fair game to him and his kind. He rose lightly, turned and stooped straight at Audax who, hearing the rush of his descent, threw himself on his back with unexpected adroitness and grabbed at the raven, who only just managed to avoid the blow. The guinea-fowl, tck-tcking frantically, made a rapid departure into their thorny stronghold.

Audax had met white-necked ravens at Oldeani and knew just what to do. Flattening out he sought the open plain and rapidly began to climb in wide falcon-like rings, rising with a speed and drive that would have been beyond the capability of any European buzzard. The raven too was no sluggard in the air and was soon climbing in smaller, tighter circles. Watching from the ground, one might have got the impression that neither bird was in the least concerned about the activities of the other. Audax was still climbing and appeared to be intent on reaching the cool greenness of the White Highlands that he loved and had so recently left behind. The raven likewise, from the speed and directness of his flight, seemed intent on reaching the Somali frontier far to the northeast. Both birds now appeared as black specks in the hard blueness of the African sky. We watched them without a word as Audax turned northward and rested momentarily on his wings before slanting across the sky in a shattering stoop that would not have disgraced a Saker falcon. We could just make out the speck that was the raven, and saw it slide-slip from the stoop as Audax hurtled past to disappear behind a barrier of scrub.

As the horizon had swallowed both contestants in this well-matched duel we never knew how it ended. The land-rover carried us to where we thought the birds had disappeared, but of course we found nothing; it is unbelievably difficult to judge distances where the whole landscape looks the same. Parking the land-rover in the shade of an acacia, we climbed onto the hood. Half an hour later we heard the well-loved AUNG, AUNG, AUNG and Audax swung into view, seeming remarkably unconcerned about the whole affair as he dropped rather nonchalantly onto the guinea-fowl lure for a well-deserved reward. He appeared to regard it all as a bit of a lark, but I doubt if the raven felt quite the same. Dingaan, who had remained behind as he disliked the heat almost as much as he disliked Audax, was far from impressed and, not to be outdone, went out and slew a puff adder. The fact that the snake was only about four inches long was beside the point; Dingaan had kept his end up, and I doubt if puff adders, whatever their size, are included in the diet of many fiscal shrikes!

Soon afterward I decided it was time to take my job more seriously and to earn my pay properly. I was determined to prospect the possibility of making at least some sort of track all the way from Makindu in the extreme north of my territory to Mananga in the extreme southwest. So with Laura loaded to capacity with food, camping equipment and water

Locust Hunter-at-Large

in every sort of container and with, I hope, the blessing of H. M. High Commission, we slipped quietly out of Makindu one morning early and headed out across the brick-red thornbush country. With the dauntless Zacharia as passenger, co-driver, navigator and indispensable factotum, and with Dingaan and Audax perched confidently atop the luggage at the back, we wallowed, grated and slid, mile after backbreaking, spring-destroying mile, heading always to the southwest.

Zacharia, a highly intelligent and adaptable Kikuyu, was a far better and more experienced driver than myself, so he would tackle the more difficult terrain, while I drove over the straightforward and more easily negotiated bits. With much pushing, straining, removing of boulders, slabs of lava, and dry termite fortresses from our path, we somehow coaxed and nursed Laura over *dongas*, dried-up river beds and countless other hazards. We each carried a *panga*, that savage, razor edged weapon-cum-implement of all work, and we hacked and hewed our way through the thorn thickets, sometimes easing our way a yard at a time along ancient rhino paths, far from certain that we would not encounter the irate owner behind the next bush. We did indeed come upon piles of steaming fresh dung and we saw more than one rhino, red-hued with the soil of its homeland, standing truculently, tail erect and horned head lifted, beneath an isolated thorn tree. But mercifully we did not encounter an ill-tempered old *faru* under conditions that would have led to a direct confrontation. Had we done so it is unlikely that I should have written this book.

Late in the afternoon we would pull up beneath the shade of a wide-spreading umbrella thorn or close to the protecting wall of some rocky outcrop that rose unexpectedly from the floor of the otherwise featureless scrubland. If possible we would press on until we came to the banks of a river that still held a modicum of flowing water. It was a curious and fascinating experience to drive on hour after hour, seldom changing out of low gear, while around us stretched the monotonous and desiccated landscape of thorn trees, withered straw-colored grass and brick-red soil and then suddenly see, far to the south, a line of living trees, yellow-green and light-fronded as feather dusters.

There was little sign of life apart from the occasional troop of pygmy mongooses and a few olive-green ground squirrels; and sometimes the silvery-gray attenuated shape of a chanting goshawk unfolding its surprisingly long narrow wings, drawing up its carrot-red legs, and dropping out of the crown of an acacia to pick up some incautious skink or dung beetle. Yet beside the river bank, yellow, black and green weavers would be building their nests in almost every tree, pied hornbills and go-away birds would honk and wheeze around us and, if we were lucky, we might catch a glimpse of a monitor—largest of Africa's lizards and first cousin to the gigantic Komodo dragon. We would drink and bathe, regardless of the alleged danger of bilharzia, refill our water containers and cook supper, which in my case used to consist almost exclusively of tinned meat, rice and thick rich gravy. I carried a twelve bore and sometimes would use it to vary this monotonous diet, for Zacharia had mastered to perfection the art of cooking francolin, guinea-fowl and white-bellied bustard.

After supper I would stroll along the river with either Audax or Dingaan, being careful to keep them apart when not supervised because Dingaan was quite ready to attack his larger and more powerful companion, and such an affray could have had only one ending. I have found that this is not an uncommon characteristic among birds of prey, and shrikes, whatever their taxonomical position, are most certainly birds of prey by instinct. Thus my shrikes were always prepared to attack kestrels or buzzards. Kestrels will often go for buzzards, and buzzards, despite their reputation for lethargy and cowardice, are by no means averse to taking on any eagle that gets in their way, or whom they may wish to rob of its prey. This particular characteristic, however courageous, makes life a bit hectic when the two would-be contestants are permanently sharing one's life, land-rover and tent.

Dingaan was a first-class companion for the sort of life I was then leading. He was small and easily transported, he was independent and could catch his own food and, above all, wherever we happened to be, he made his home for as long as we remained. He showed little propensity to wander, came when called as promptly as a well-trained dog, and seemed to enjoy his nomadic life to the full. Audax on the other hand was more inclined to roam, never under any circumstances came when he was called, and only condescended to accept the lure when he felt in an unusually cooperative mood. Thus, despite a few brilliant exhibitions, he was by no means a first-class hunting bird, being lazy and placid. His latent fire only showed if he was exceptionally sharp set or if his strongly developed territorial instinct had been challenged by the appearance of some other raptor, especially the intrusion of some member of the crow family.

Time these days meant nothing; I seldom had any idea as to the time of day, the day of the week, or even the month. I would get up early, make coffee, boil an egg or two (guineafowl's or bustard's for preference), help load up Laura and we would be on our way, navigating by the sun, eating when we were hungry and sleeping when we felt like it. We had no one to tell us what to do or where to go, and all around us was the great pulsating heart of Africa—untamed, unexploited and as it had been ten thousand years before. It was a strange and a satisfactory feeling to realize that here were two men, one European and one African, two birds of totally different species and temperament, and the tiny dot that was our landrover, moving together through the savage landscape that stretched mile after unchanging mile, all around us.

Day after day we pressed on, rolling and rocking through the unending bush; how many times we had to stop and repair our tires, lacerated by the bayonet-like thorns, I shall never know. Our water ran out and we had no river to help us but we had, with this sort of eventuality in mind, filled all sorts of containers with Coke or Pepsi Cola and this sweet and somewhat unrefreshing beverage nonetheless kept us going. Zacharia and I, despite our totally different backgrounds and approach to life, had become remarkably close during this trip, even though my Ki Swahili was execrable and his English was not much better. But Zacharia had courage, ingenuity and a sense of humor. He did not have to be there; he could easily have found better paying and far less hazardous employment on any European-owned *shamba* in the Nairobi area. Nevertheless he had thrown in his lot with mine and he never once complained other than to curse the evil spirits that somehow contrived to put so many obstacles in our path.

When Coke ran out we were reduced to sucking pebbles and chewing the horny outer skins of some extremely lethal-looking fruits that even the desert-haunting birds eschewed. Dingaan and Audax seemed to thrive on the moisture that they must have found in the insects, lizards and small rodents they accounted for. We came across and dealt with a small concentration of locust hoppers, having piled in a few sacks of Gamaxene-treated bran before starting out. I knew that the Africans occasionally ate locusts, and that nearly all African animal life, from mongoose to ostrich, with any pretensions to being carnivorous, will snap them up whenever they get the chance. I had heard them described by Europeans who had been rash enough to try them as tasting rather like shrimps so, when shortly afterward we ran into a small resting swarm of adults, I decided to sample a few myself. Zacharia filled a frying pan with fat and locusts and did his worst! I can now say from experience that locusts do not taste in the least like shrimps; in fact I found them totally inedible.

By now I was pretty light-headed and nothing seemed to matter much. I had brought a bottle of "Beehive" brandy in case of emergency, and it seemed as good a time as any to tap it. Never has brandy tasted better or its effects been more welcome. Zacharia preferred *pombe* or home-brewed native beer, and had probably never tasted spirits before; but now he took his tot. Thus physically and mentally refreshed we would have driven to hell and back, and hell itself could hardly have been hotter or drier.

All the while, slowly but surely, Laura—aided, abetted and sworn at—crawled southward with dogged determination. She did her manufacturers proud: seldom could a vehicle have been put to such a test and battled so gallantly. What Zacharia's thoughts were I could only guess, but I was gripped by a curious sort of exaltation, a feeling almost of unreality, and never for a moment did it occur to me that we might possibly fail to get through. At last, after forcing our way between a barrier of tangled bush and a vertical wall of red earth and lurching down a steep boulder-strewn slope, we found ourselves on what looked like a man-made track leading, by all that was wonderful, in the direction in which we were heading. Imprinted in the silvery dust were the marks of many vehicles.

We drew onto the side and halted; Zacharia and I shook hands. *Mazuri kabisa* (very good indeed) was all he said. I had a therapeutic tot of brandy, and Zacharia joined me. We still had half a kerosene tin of warm dirty water for the land-rover, after all she couldn't drink Coke and brandy—though on reflection I feel that she would have done so on this occasion. We rested a little while Dingaan, unruffled as ever, hunted grasshoppers and Audax gave himself a thorough dust bath, shuffling about on his breast, flicking his expanded tail to let each tiny particle trickle through his plumage to his skin. We lay dozing and relaxing until dusk, when we boarded the landrover once more and, moving for the first time for days with satisfactory speed, were soon devouring the miles.

Dusk, as always in Africa, slipped rapidly and unobtrusively into full darkness, and from somewhere in the south appeared the glow of the rising moon. But as we traveled on the whole horizon appeared illuminated, as if some large town lay just ahead; it suddenly struck me that it could not be the moon after all. The weird glow intensified as the track gave way to open dusty plain. Speeding across it, drawn like moths from darkness to the light, we reached a half circle of familiar flat-topped thorn trees among which loomed strange bulky shapes, and from surrounding trees were slung what looked like and proved to be arc lamps. Drawing up close, the black elephantine shapes were now disclosed as marquees. Surely the trials and privations of the journey could not be having hallucinatory effects on me? These could hardly be the figments of an overwrought mind; and besides Zacharia saw them too!

I got out of the land-rover and took Audax on my fist. A light gleamed through the canvas of the nearest, smallest tent. I hesitated, took a deep breath and rudely shouldered the canvas flap aside. A figure rose from a wicker chair, came forward to meet me and said, "Well, well, come in and make yourself at home!" It turned out to be Nancy Russell-Smith, my climbing companion in the ascent of Oldonyoroc. In another chair sat a man in bush jacket and shorts, and in between, I was thankful to see, was a table with glasses and a bottle. All I noticed about the man was that he had a monkey on his shoulder. This was not unusual, because in East Africa at that time many people carried monkeys. What was less usual was that this particular simian was not the usual monkey.

9

The Rhino and the Baboon

FIXING ZACHARIA UP for the night in the comparative luxury of the quarters reserved for African employees, I returned to the tent. It appeared that we had blundered straight into the very heart of Filmland. Here, in the wilds of savage Africa, Ealing Studios had arrived with a full company to shoot the film *Where No Vultures Fly* in the Amboseli National Park. Zacharia and I, rolling and swaying our slow tortuous way in from the northeast, had never heard of them or their much-publicized arrival. Making my camp near by, I later had a wonderful bath, perhaps the most welcome of my life, provided by the good offices of Ealing Studios. Later, clean, cool and comparatively tidy, I rejoined Nancy for a welcome reunion chat, who as official caterer, adviser and general Den Mother to the company, was thoroughly enjoying herself.

The next morning, after one of the best nights I can remember, I rose late to find Zacharia equally refreshed and as tidy as if he were still employed by the Harrisons in Nairobi. I had decided that we could do with a few days' holiday and Zacharia, enjoying the new experience nearly as much as I was, seemed delighted. Nancy knew everyone and in the course of the next few days she introduced me to everyone who had anything to do with the production, from Leslie Norman, the producer, to Anthony Steele and Diana Sheridan, the two principal stars, and to William Simon, the juvenile lead, who in my opinion made the whole film worth producing. William had already introduced himself to me, or rather to Audax, whom he met perched outside my tent. I later came to know him and his mother quite well and found them the most pleasant and unaffected people. William had adopted a Thomson's gazelle fawn, which he called Lucky, and whom I met again a year later at the London Zoo. As the film was about a wicked game poacher and a courageous, honest, upright game warden, the whole place was swarming with animal extras, ranging from a family of young barn owls to Iola, a tame lioness, who took a small but important part in the story.

That was where Sally, the baboon, came in. An orphan, whose mother had been killed by a leopard, she had been found by Sandy McGillivray, a road foreman for the Public Works Department. How he had become mixed up in the film escapes my memory, but there he and Sally were. Until I met Sally I held much the same views about baboons as most residents in East Africa, who regard them as ill-tempered, obscene and dangerous brutes, with a penchant for murdering young lambs in order to obtain the milk in their stomachs. There is something vaguely sinister and repellent in the very name that conjures up pictures of the sort of misshapen, vindictive, subhuman creatures that I associate with Dante's *Inferno*. Thus I was not prepared to go overboard for Sally's charms. But Sally herself had missed her vocation in the Baboon Diplomatic Corps, as an ambassadress at least.

A day or so after arriving at Amboseli I was exercising Audax and Dingaan near my camp, a good way from the film set. Amboseli for the time being bore a close resemblance to

The Rhino and the Baboon

what I imagined Hollywood to be. There were people everywhere—impressive-looking men in loud shirts, dark glasses and long-peaked caps abounded, though whether they were directors, producers or just clapper boys I never did discover. Then Sandy appeared with Sally scampering at his heels like a puppy. In fact her whole appearance had a strangely canine look about it. She had a dog-like head and a shortish hairy tail that she carried either bolt upright or curled over her back, according to her mood. She walked slowly on all fours when she wasn't standing up, with her small hairy black hand grasping Sandy's large, pink and horny one. She must have been between two and three months old and her quiet dignity, alternating with youthful boisterousness, quickly warmed me.

Sandy sank into a safari chair and Sally clambered onto his knee, peering at me with interest from her deep-set hazel eyes as I poured cups of coffee. Zacharia brought in some biscuits and I offered Sally one, who reached out a lackadaisical paw, took it delicately, and bit a piece off. Liking the taste, she grunted her thanks; Sally always grunted when she was pleased. I gave her a cup of diluted condensed milk. She took the cup in both hands and sipped the contents. When she had finished she descended from Sandy's knee and lolloped across to me, moving on three "legs" and carrying the cup in one hand. Handing it politely to me for a refill, she climbed onto my knee and studied me carefully between sips. Having finished the second cup she smacked her lips loudly to show that she was satisfied and belched gratefully. Then she took one of my hands in both her front paws, carried it to her mouth and very gently nibbled my thumb. That was Sally's way of kissing. Dingaan was grubbing about the floor looking for beetles when Sally dropped to the floor to examine him. Dingaan, who must have had a hereditary knowledge of babboons and their wicked ways, shot into a tree and let rip with every oath in the shrike vocabulary—and Dingaan could really use 'orrible language when he felt that his dignity had been outraged. Sally listened to this disgraceful performance with an air of slightly pained concentration which her beetling brows and close-set eyes enhanced.

Sandy, it seemed, was due for home leave and was desperately anxious to find a kind and understanding home for Sally, to whom he was as devoted as she was to him. Unfortunately no one seemed anxious to saddle himself with such a companion and Sally herself was by no means universal in her affections. She and I, however, had by now developed a sort of rapport and I was fascinated by her, never before having had close contact with such a beast. Monkeys were about the only animals that had never appealed to me, but Sally was no ordinary monkey. Her manners were normally beyond reproach and she comported herself with a quiet dignity unlike the nervous restlessness, combined with insensate mischievousness, that was so much a part of the ordinary monkey. On reflection I would say that the baboon (at least the olive baboon of East Africa) holds much the same position among the primates as the raven does among the other crows.

When Sandy mentioned his predicament to me one evening over a sundowner, without hesitation I offered to take her on. I have seldom seen anybody so relieved. Next morning I fastened a cord to the leather strap around her loins and tied one end to a tent peg while I watched Sandy and his landrover disappear toward Mananga in a smokescreen of silver dust. When she found that Sandy, her friend and protector, had gone and did not return, Sally's grief was tragic. She put her hands over her eyes and retired to her own specially made wooden cradle-cum-traveling box, wrapped herself up in her blanket, and refused to be comforted. She neither ate nor drank but just lay beneath her covering, a tiny quivering heap of misery. I tried to cheer her without much effect, although she did catch hold of my hand with hers and clung to it

The Rhino and the Baboon

desperately until she seemed to have fallen asleep. I had a good deal of difficulty unhooking her tiny clutching fingers without waking her, although she must have been worn out with grief, a feeling I remember only too well from similar situations in my early childhood.

I knew that if there was one thing that Sally loved other than Sandy McGillivray it was a cup of warm sweet tea, the sort we used to refer to in the army as "sergeant-major's tea." When that afternoon I returned to the tent, a small rotund lump, hidden under a rather sad-looking tartan rug, showed where Sally slumbered on. I asked Zacharia to bring two cups of hot, strong tea, and a tin of condensed milk. I poured myself a cup and made as much noise as possible, clattering the cup and saucer and banging the spoon on the table. As I glanced at the cradle, I noticed a slight disturbance, then a small lugubrious face appeared. Sally had taken the hint. I took a lump of sugar, well soaked in the nauseatingly sweet concoction, and gave it to her. She took it in one hand, sniffed it and crunched it up. Somehow she was out of bed and on my knee in one scrambling bound. She took a cup of tea in both hands, leaned her back against my chest, and began to drink it with obvious satisfaction and much smacking of lips. Having finished it, she handed it back and indicated that she could do with another. She picked up a piece of fruit cake delicately with her lips and finished it to the last crumb. She had now obviously accepted me as a substitute, albeit a poor one, for her original friend and foster father, and she climbed onto my shoulder where she felt secure and from which elevated position she could survey the surrounding countryside. Later, after Sally had been my companion for some time and we had grown used to each other's ways, I suspect she regarded me rather unflatteringly as a substitute for the big male baboon, leader of what should have been her family pack. If she was in any way scared or upset when out with me she would come scurrying up and swing onto my shoulder and if possible my head, which she would clutch tightly with all four feet, grabbing my hair as an additional support. This was a manifestation of trust and affection which I did little to encourage.

A few days after Sally had been handed over to me I had to drive to Mananga to get some supplies, and I took her along in the land-rover. I also took Dingaan but Audax remained behind on his block under the shadow of a tree close to my tent with Zacharia to keep an eye on him. Sally must have traveled many hundreds of miles in her short life. Sandy McGillivray had been in charge of road construction and repairs between Arusha and Dodoma, a distance of two hundred miles or more and from the age of six weeks or so Sally had gone with him as he moved about his business. A perfect passenger, she would sit on her haunches in the passenger seat (or on the knees of any human being who happened to be there), gazing out of the window, apparently lost in her own thoughts. If, however, she saw lion or cheetah or any other potentially predatory enemy she would bob up and down barking loudly, at the same time making appalling grimaces at the totally indifferent foe. She obviously felt secure and free from any danger of attack. If, on the other hand, she should encounter a similar threat while out walking with me (and both lion and cheetah, not to mention her chief hereditary enemy, the leopard, were by no means uncommon in that area), she would rush up to me, leap aboard, and sit silent and trembling, holding on tightly until we had retreated to what we considered a safe distance, whereupon she would drop to the ground, strike the earth with the flat of her hands, hurl unprintable abuse and challenge the now distant predators to stand up and fight like a man. She was, in short, an endearing and highly individual companion.

On our first drive together I called at the local *duka* (whose owner, incidentally, must have been coining a tidy fortune from the film company) and stocked up with all the essential things that one only notices when one has forgotten to buy them. Then after a long cool drink of pure lime and well-chilled water that Sally enjoyed as much as I did, we went for a stroll along the banks of the Mananga River. It was a long time since I had been able to walk in real, thick shady forest and hear the chuckling and gurgling of fast flowing river water and I enjoyed it to the full, as indeed did Dingaan and Sally. Dingaan lost no time in flitting to the nearest hollow where the water formed a shallow saucer with a handy rotten branch lying half submerged across it. Here he gave himself a really satisfying bath, soaking his body until he was incapable of flight and had to be lifted to an overhead creeper cable to dry off.

Sally cupped the translucent water in her hands and drank deeply, after which she gave a good display of her latent baboon nature, turning over flat stones to look for scorpions, centipedes and other loathsome forms of primitive animal life which she snapped up with suitably appreciative lip smacking. I noticed that she knew just how to deal with a scorpion although I doubt if she had had any previous experience. On disclosing the hiding place of one of these furtive arachnids she would feint with one paw until its segmented tail, with its venomous thorn-shaped sting, whipped forward over its back in a threatening gesture. Then, quicker than light, she would seize the tail from behind with the other hand, rendering it harmless. This she would break off and cast aside, much as a gourmet discards the inedible portion of a prawn. She would treat the crab-like pincers in similar fashion, after which the body, or what remained, would be eaten with supreme enjoyment. It occurred to me, watching Sally systematically up-ending stones, that a troop of baboons must do an enormous amount of good in destroying not only scorpions but tarantulas and other small obnoxious beasts.

As I crossed the stream by a series of natural stepping stones to examine some unusual Skipper-like butterflies that were playing tag about a tangle of sun-embraced creepers just ahead of me, I heard an indignant scolding bark behind me, half-pleading, half-condemning. Sally, who had followed me into midstream, was teetering on a stone that threatened to topple over and decant her into several feet of swirling water. Her look of horror was pathetic but at the same time almost comical in its so nearly human expression. I offered her my hand which she took quite calmly, and as I swung her up and over, plonking her on terra firma on the opposite bank, it was exactly like lifting a little girl over a puddle. While I was on my entomological inspection Sally continued her hunting with inexhaustible zeal. It was amusing to watch her plodding flat-footed along the bank, her tail curved over her back and cocked to one side, her expression one of intense concentration as she sought out her revolting appetizers. She looked exactly as if she were posing as the big fierce grown-up baboon she would no doubt one day become.

Unexpectedly her mood changed and she leaped for the lower branches of a small tree, and with amazing dexterity and speed swarmed up to the top where she performed a sort of triumphant hornpipe. When I called her she came down almost at once, flinging herself into my arms from a branch several feet above my head. It was a most endearing display of trust and affection. Later, when we knew each other better, I found that she would always let herself go, falling from surprising heights, knowing full well that I would always be there to catch her although, as she grew in weight and stature, this delightful trick could prove a bit disconcerting. On my return across the stream she rode pick-a-back, as if I were her mother. Collecting Dingaan, now restored to his usual spruceness of feather, I returned to Amboseli with my supplies.

The Rhino and the Baboon

After this Sally went nearly everywhere with me, taking the place of a dog, which in certain ways she resembled, being just as affectionate, loyal and willing to please, but with a pronounced sense of humor lacking in most dogs. Unlike a dog, however, she could climb trees, from which after much patient training I taught her to retrieve certain fruits such as guavas, mangoes and paw-paws. She was a good guard too, viewing strangers with considerable suspicion; and she was sufficiently formidable in appearance to deter any potential evildoer who might have been tempted to rob my tent or car. I would leave her in the land-rover with Audax, while I went about my business, and although I frequently had considerable sums of money concealed in the vehicle I knew full well that Sally and Audax were redoubtable enough to ensure that my possessions remained unrifled even at the height of the Mau Mau disturbance.

One day in Arusha, after a particularly grueling safari, I was headed for a hotel for a much-needed wash and brush-up, with Sally ambling beside me, when as luck would have it I ran into an old acquaintance of mine, a particularly debonair and immaculate ex-cavalry major with whom I exchanged a few pleasantries. Years later, in a London club when I met him again he gazed doubtfully at me for a few minutes and remarked, "Ah yes, I remember you. The last time we met you were walking down the main street of Arusha arm-in-arm with a gorilla. The gorilla needed a shave and so did you!"

About this time I decided to pay a flying visit to Oldeani and look up my friends the Kiellands with whom I had kept in infrequent touch. I knew I had a standing invitation there, and besides I had received news of a locust swarm in the area. The fact that it was about two hundred miles outside my district did not concern me one iota. It was an eventful visit.

The second morning after my arrival Audax took off after his old enemies, the white-necked ravens that nested in the ravine below the rondavel in which I stayed. After a spectacular display of stooping and evasion on both sides the three contestants rose to such a height as to appear like wind-tossed leaves as they drifted away toward the distant M'Bulu hills.

By nightfall Audax had not returned, though the falsetto croaking below on the cliff told me that his opponents were back at base and were no doubt discussing stoop by stoop the action of the day and their valor and prowess in outwitting such a formidable foe. The following day, despite much calling and whistling, there was neither sound nor sight of Audax and the sky, the forest and the valley below seemed the poorer for his absence. I was not worried for his safety. I knew that he could catch his own food and I also knew that his jesses were much too short to catch in any natural projection. I was upset on purely selfish grounds, being fond and proud of Audax and enjoying him around. That evening I was sitting with Sally, watching the sun slowly descending over the escarpment that guarded the approach to the Ngoro-ngoro crater. While listening to the clamor and admiring the speed and aerial dexterity of a pair of white-rumped swifts, I heard a familiar ringing call and saw Audax sweeping in, traveling low and fast from the direction of the darkening forest behind me. He swung up onto the peak of the rondavel where he sat, looking like an extremely handsome weather vane, until I called him down for a meal that he ate without enthusiasm; his mind was clearly elsewhere.

I did not have long to wait. From the fringe of the forest came a loud demanding call, like Audax's own but deeper and more authoritative. Audax cocked an eye in the direction of the call, but went on feeding absentmindedly while from the dusky forest belt floated a darker shadow that circled out over the Korongwa, spiraling in wide, lazy circles. It was a big black female augur buzzard, whose fox-red tail was the only relieving feature in her otherwise somber plumage. I had seen this solitary melanistic female many times before when I was staying at Kit'Ndovu and was surprised that she had no mate. Adult augur buzzards usually keep together in pairs—hunting, playing aerial games and roosting together at night. The dark color phase is by no means uncommon among these handsome birds. Now at last, it seemed, she had laid claim to Audax's heart, a demand far stronger and more natural than anything I had to offer. Audax looked at me, wiped his beak on my finger, roused his feathers once and launched off to join his newfound mate. As he turned in his spiraling ascent, the fast dwindling sunlight caught his snow-white breast and turned it into a delicate shade of pink as the two birds climbed together to disappear behind the forest canopy in the direction of the Oldeani crater.

I sat on deep in thought, until Sally, with a confidential cough, climbed into my arms and snuggled close against my chest. I carried her into the hut where Zacharia had lit the pressure lamp. Dingaan was already there, perched upon his chosen vantage point, one of a pair of ancient buffalo horns—huge, black and menacing—that loomed from a wooden plaque above the door. Dingaan, I am convinced, was quite certain that he had driven Audax away, and consequently his arrogance and self-assurance were redoubled. He would fly to the top of the rondavel, once his rival's favorite look-out, to descend in one plunging stoop onto any grasshopper, beetle or small rodent that was rash or unlucky enough to catch his ever watchful, ever rapacious eye. He would alternate these forays with snatches of song.

This new-found arrogance was very nearly his undoing for one evening, as he flaunted his voice and flirted his tail aloft, an arrowy shape silently and swiftly dropped upon him from a nearby tree. Quick as was the aggressor Dingaan, despite his tameness and lack of experience, was faster still. With a rasping challenge he dropped from the rooftop and half flew, half rolled into the doorway of my hut as the gray marauder grappled with him. Only half aware of what was happening I dashed to the rondavel, stepped inside and slammed the door behind me. In a corner beneath my camp bed I could just make out two struggling figures. Beneath, on his back, and fighting desperately with beak and powerful feet was Dingaan while above, holding him down, was one of Africa's smallest and rarest birds of prey, the East African little sparrow hawk, only slightly larger, but more powerfully armed than her intended quarry. She had locked her talons with Dingaan's claws and so for the vital few seconds before I arrived had been unable to inflict mortal damage. I reached under the bed and grabbed the combatants by whatever part of their anatomy I could reach. The little sparrow hawk tried to release her hold but Dingaan's black feet were still interlocked with her yellow ones. Her brilliant orange eyes blazed into mine with splendid savagery and defiance, and she opened her beak and sent forth a ringing falcon-like call of fear and anger; freeing one foot she clutched my finger, making me wince with the unexpected power of her grip. I managed to separate the pair at last and holding the little hawk in the way one would hold a racing pigeon, took it to the door and tossed it into the air. Hurtling into the gloom, she was instantly swallowed up by the purple shadows. As a result of this contest Dingaan lost two tail feathers, one primary and his dignity, and he took a few days to return to his normal ebullience.

Almost immediately after the little hawk melted into the dark shrouded forest I cursed myself for releasing it so quixotically. Here in my hand I had held a rare and exquisite creature, about which little appeared to be known. Few, if any, falconers ever had a chance to train and fly the little sparrow

hawk and I never had another. On reflection I consoled myself that she might have had a family of eyasses to look after, that she was old and past her best (she didn't look or behave like it!), and in any case she would probably have died from one of the many diseases or accidents to which wild hawks are so prone when in captivity. Dingaan's point of view wasn't considered, but there is little doubt that she would have killed him in the end.

It was time I went back to Kajiado, to the open windswept veldt land of the Arthi Plains and to Nairobi where no doubt the Chief Locust Officer would want to hear what I had done to justify my princely salary in the last few weeks. On the last morning I was to spend at Kit'Ndovu I decided to take a final look at my beloved rain forest behind the *shamba*. I left Sally with the Kiellands, who had become very fond of her in the short time they had known each other, and in whose company she was perfectly relaxed and at ease. Sally and Margit's little daughter, Anne-Elise, had quickly reached a mutual understanding and were a delight to watch as they went about their own affairs together.

I had always liked to be alone in the forest; somehow the human voice, however muted, seemed as out of place there as in a great cathedral. The voices of nature were so utterly different, so completely a part of the unspoiled, unexploited wilderness, unchanged since time began or so it seemed to me as I pushed further into the welcoming greenery. As I sped along the game trails I remembered so well, all around me were the voices of the forest. Blue-gray Sykes' monkeys crashed and hooted as they followed me, keeping to treeways running parallel to my path, inquisitive to see what I was up to and unflattering in their comments on my appearance and probable intention. Touracos uttered their pheasant-like crowing calls and hornbills honked and brayed, sometimes sailing across the space ahead of me, their lazy flapping flight and prehistoric casqued beaks adding much to the atmosphere of untamed primeval solitude. I felt then, as I have felt so often before and since, that I was the only human being in the whole world—stretching ahead mile after countless mile of forest, plain, and mountain range, with never a city, town or settlement of men to mar its unspoiled beauty.

The curious, almost equine, odor of rhino was strong in my nostrils as I hurried along the track, beaten and trampled flat by the feet of generations of these huge, peaceful animals who for centuries had used these labyrinthine ways. I knew that, provided I was circumspect and did not provoke trouble, I was safe. The chance of another unprovoked attack by an unwounded buffalo was so unlikely as to be unthinkable but the remote possibility added just the right amount of tension to make my journey that little bit more worthwhile.

A huge, dark bird, larger than any eagle I had ever seen, dropped out of a great tree ahead of me and floated on owl-silent wings to disappear where the track veered sharply and for no apparent reason to the left. This was my first sight of a crowned hawk eagle in the wild. I was as excited as if it had been an okapi and hurried on in the faint hope of another and clearer sighting of the largest and most powerful of all of Africa's raptorial birds seen, as it should be, deep in its own ancestral forest home. Searching the treetops above me I followed the track and all but plunged into the pitfall game trap that gaped open-mouthed at my feet. I knew that game poachers used traps of all kinds and I also knew that a friend of mine at the Egerton School of Agriculture had been very nearly strangled in a snare set for bush buck high up in the Aberdare Forests. But I had never seen one myself and I became exceedingly angry as I contemplated the man-made atrocity in front of me. What made it worse was that it had claimed its victim. In the bottom of the trap, about six feet down, was a small and irate rhinoceros calf; how long it had been there was anybody's guess, but mercifully it appeared to be uninjured and not particularly emaciated.

I looked furtively around. I had no wish whatsoever to encounter "Mama" or her well-justified wrath at the moment. Knowing that cow rhinos seldom desert their calves, I felt certain she must be lurking somewhere close at hand. Rhino spoor was everywhere, but then it was a rhino thoroughfare. My return, as I jog-trotted along the track back to the farm seemed endless. I had no idea I had traveled so far. I turned the final corner at last and came out into the sunlit, smiling clearing that was the *shamba*, and soon reached the farm where the entire Kielland family was now assembled. Within minutes a gang of African estate workers, with ropes, poles and *pangas*, were on their way to the rescue. Jan strode ahead with his heavy rifle slung over his shoulder, just in case it should be needed.

At the trap all was unchanged. The mother must have been forced, by hunger, thirst or some other cause, to move away, leaving her calf to its fate. She, poor thing, could in any case have done little to effect a rescue. The baby rhino, despite its ordeal, was full of fire and fury, striking out left and right with its tiny incipient horn at any rope that was lowered into the pit. Eventually a loop was successfully angled over the calf's head, thus to some extent restricting its movement. Two Africans scrambled down into the pit and after some difficulty another rope was passed beneath the animal's body and out on the further side. After much heaving and sweating, the young rhino was hauled to the surface and with his four legs immobilized with rope was lashed to a sort of litter made out of poles. Other workers lost no time in filling in the deadfall; this one

The Rhino and the Baboon

at least had claimed its last victim. Then, with the Africans chanting lustily as is their way after a successful venture, we set off for home.

The little rhino was only the size of a warthog. Borne on the powerful shoulders of eight Umbulu tribesmen he was soon on his way to his new, if temporary, home. By a lucky chance there was an empty cattle *boma* normally used for housing sick or injured oxen until the arrival of the vet, and our hefty foundling was installed therein. On being freed from the rawhide thongs that had held him captive he immediately charged at his well-wishers, scattering them like rooks before the peregrine's stoop. It was obvious that he must be fed without delay; he couldn't have been more than three or four months old and was entirely unweaned.

At Kit'Ndovu, as on most East African farms, the estate workers were able and indeed encouraged to grow a great deal of their own food; they were also permitted to keep a herd of skinny, but exceptionally hardy, cattle of Zebu type and also a herd of goats. Their goats were enormous beasts, with drooping ears, backward slanting horns and thick shaggy coats which, despite the midday heat, did not appear to cause them any undue discomfort. They were adventurous wanderers, completely fearless, and indeed had much of the appearance and a good deal of the temperament of Afghan hounds. Now it happened that the largest and most self-assertive of the nannies had only a day or so previously been deprived of her kid by a leopard or some other predator, and her udder was swollen and uncomfortable, despite the efforts of the herdboy to relieve her. What more promising foster mother for our youthful but already substantial waif could there be? The old *n'busi* was rounded up and led protesting with both voice and horns to the cow byre where the little rhino was waiting thirstily. She was shuffled inside and her makeshift halter removed. Now came the crucial moment: would two such totally different beasts accept each other and, if so, assuage each other's needs?

Their first reactions were not promising. It was hard to say which of the two was the more horrified at the other's appearance. Even to an animal fanatic such as myself their looks left a great deal to be desired. It was twilight in the *boma* and we could just make out the gray shape of our guest standing apathetically in a corner. On seeing the light from the door and hearing our entry he shambled forward hopefully, uttering a series of pathetic squeaky bleats. When he caught sight of the hairy apparition we had produced, he stopped dead in his tracks and a look of such disappointment swept over his already forlorn features that it would have been almost comical had it not been so sad. However, mingled with but not quite submerged by the rich, ripe and well-seasoned pong of goat, was another odor, one he knew well and which his lonely soul craved, the aroma of warm, fresh milk. Besides, the goat for all her unprepossessing appearance was a four-legged beast like himself and, furthermore, unquestionably female. The goat, for her part, despite her almost mythological appearance, was warm-hearted and yearned for something young and helpless to take the place of that which she had so recently lost.

The herdboy drew off enough milk to fill the hollow of one hand, and we cornered the rhino who was still uncertain as to our intentions. I took a few drops of the milk and rubbed it on the orphan's prehensile lips. The magic worked; a long questing tongue came out and licked the frothy liquid till not a drop remained. We then urged the goat forward, still a little apprehensive, and somehow managed to insert a teat into the calf's mouth. After a few abortive and disappointing trials he suddenly got the hang of the thing and began to suck. We stole out quietly closing the door, leaving the two together. An hour or so later I peeped into the shed and in the dusky in-

terior I could just make out a heart-warming sight. The rhino calf, reclining on the thick straw, was sleeping the sleep of happy repletion, while the old nanny goat was caressing his leathery mud-caked flank with her long pink tongue. Hearing the door open she turned toward it, stamping a cloven foot and shaking her head menacingly at me. No one was going to come between her and her foster son if she could help it.

Not far away on the other side of Oldeani settlement lived a retired Afrikaans white hunter who many years before had given up the rifle and become a devoted champion of African game animals large and small. That evening we drove over to see him and explain our position. Jan was far too busy running his farm to take on such a responsibility, while I was always on the move and a rapidly growing rhino was, however appealing, hardly the ideal traveling companion for one of enforced nomadic ways. Besides, even Sally and Dingaan took a lot of coping with. Van Royen was delighted; he was particularly fond of rhinos, which he regarded as nature's gentlemen—a bit choleric at times perhaps, but simple-hearted, courageous and conservative in habit, asking only to be left alone to mind their own business and expecting everyone else to mind theirs.

10

A Home for Karen

THE NEXT MORNING I left for my district. I had agreed to deliver the rhino and its foster-mother en route if this could be arranged. After a delicious breakfast of maizemeal porridge, grilled buffalo kidneys and home-grown coffee, we started loading up for my return journey. The goat was led out of the *boma* with Rocky the rhino, now completely at home and a good deal tamer than the average pony foal, cavorting along behind. After a battle of rodeo-like proportion we managed to hoist the old "she-satyr" into the land-rover, where she was secured by one of the farm boys who was to go with us as far as Van Royen's *shamba*. Now it was Rocky's turn, and though we prepared ourselves for a desperate duel in the sun, our anxiety was unfounded; Rocky had no intention of being abandoned again and hearing his foster-mother bleating from the interior of the land-rover, made frantic attempts to join her. A rough-and-ready ramp was prepared, up which Rocky clambered into the back of the vehicle with the aplomb of a well-trained and much-traveled dog. Sally, who had been sitting on her haunches, chin in hand, watching this performance with the rapture of a child at the circus, climbed into the passenger seat where Zacharia was already ensconced. Finally, Dingaan was whistled up and shut safely in his traveling box.

We shook hands all round, and after much mutual well-wishing we were off, negotiating the well-remembered but nonetheless awe-inspiring corkscrew descent, with its terrifying view on our right of the sheer drop down into the lion-colored, lion-haunted scrubland thousands of feet below. Throughout that hot, dusty and thoroughly uncomfortable journey both Rocky and his foster-mother behaved impeccably; nevertheless it was with considerable relief that I saw the blue gum trees that lined the drive leading to Van Royen's estate. A splendid fellow, Van Royen must have been well over seventy at that time, and might have been the prototype of one of John Buchan's fictional heroes. Calm, kindly and courteous, he took charge of the operation. Without fuss Rocky and the nanny goat were led away to a large comfortable stockade, already occupied by two zebra foals and a delightfully tame and friendly cow eland who, incidentally, conned me out of most of my safari ration of cheese sandwiches, for which it appeared she had a passion.

Van Royen was accompanied on his rounds by a litter of three-month-old wild dog puppies which, with their bat ears, pot-bellies and hyena-like hind quarters, belied the reputation for speed, savagery and limitless stamina with which they are credited. I had always believed that, like the Scottish wildcat, these curious carnivores, which are not true dogs at all, were untamable from birth. But these puppies were as friendly and frolicsome as any domestic canine. Pride of place however in Van Royen's eyes was occupied by Rajiki, a two-year-old male baboon of the same species and acquired in just the same way as Sally. Rajiki (which means chum or mate in Swahili) was a formidable beast with colossal canine teeth and, though far from full grown, gave promise of developing into a force

A Home for Karen

to be reckoned with. He treated Sally with splendid contempt and Sally, shy, bashful and her nose well out of joint for the first time since we had met, refused to leave the land-rover. However, we had business to discuss and coffee to drink; an hour or so later we descended from the stoep of the house and were astonished to see that Sally and Rajiki, left to their own devices and free from parental interference had, like two children, not only made friends but were quite happily engaged in demolishing the place, splinter by splinter and stone by stone.

It was quite clear that with the influence of the stronger-willed Rajiki, Sally could easily become a proper little "mafiosa." Van Royen was as enchanted with her as was Rajiki and made none too veiled hints that, should I have to part with her at any time for any reason, he would be only too pleased to welcome her into his already diverse family. I was relieved to hear this, for I was vaguely planning a holiday in Europe as soon as I could accumulate a sufficiently well-nourished bank balance (which, as always, appeared a pretty remote possibility but worth pondering nonetheless), and the prospect of arriving in chilly, rainy, fog-bedeviled Britain with an animal as susceptible to cold and its accompanying ailments as a baby baboon was unthinkable. This, however, was in the most unimaginable future. Meanwhile the two miscreants were happily employed impersonating Hilary and Tensing on a mountain of newly made bricks in the center of the farmyard, sending them slithering to the ground in an avalanche of red dust. Although destruction was appalling, Van Royen seemed completely unperturbed. I collected Sally, who seemed none too pleased to leave her new playmate. Nevertheless, clutching half a brick as a souvenir, she was finally installed with Zacharia in the passenger seat, and we were on our way without further incident, pausing briefly in Arusha to warn the game warden of the activities of poachers in the Oldeani area.

A few days after my return to Kajiado, as I sat in the shade of a great wide-branched umbrella tree writing out an embarrassingly overdue report on recent happenings, I was approached by Zacharia. He coughed apologetically and stood to attention (he had served in the Kings African Rifles for a short time and this was always a sign that he was about to ask a favor). He asked me if he could have a few days' leave, which I gladly granted him. I inquired casually if he had any particular plans and he replied that he wanted to buy a wife, which seemed reasonable as he did not appear to have one at that time. Thanking me, he went off on the Nairobi-bound bus. A week or so later he duly returned and presented himself for duty. I asked him whether he was now happily married and what, incidentally, he had done with his new bride. A bit noncommital, he merely told me that he had installed his wife in the family *shamba* at Kiambu, outside Nairobi. For the next few weeks, unless I had a special job for him to do, he would go home at weekends, leaving on the Friday and returning on Sunday night or Monday morning.

One day I had to go into Nairobi on locust business and dropped Zacharia off, telling him to meet me outside the Mansion House at a certain hour that evening. He was there on time, talking to a Kikuyu girl who was, to put it mildly, remarkably uncomely. I asked him if he had had a satisfactory day, and he seemed somewhat uncommunicative, which was unlike Zacharia, normally a cheerful and highly vocal fellow. However I forgot all about this until a few days later, when he came to me and asked for a few days leave.

"But, Zacharia," I said, "You've only just come back from leave. What on earth do you want now?"

"I know, *effendi*," he replied with downcast eyes, "but this time it is important."

A Home for Karen

"Surely, buying a wife was important, Zacharia?" said I.

"Ah, *bwana*," said he. "This time I go to swap my wife for a fine bicycle!"

Having seen the lady in question I could only sympathize and let him go without further ado. The bike turned out to be a beauty, an almost brand-new Raleigh and I have seldom seen Zacharia so proud and happy.

Not long after this I received a message that Dr. Leakey of the Coryndon Museum in Nairobi had been presented with an injured hawk, and as he did not want it as a museum specimen, I was welcome to it if I would drive over and pick it up. Within half an hour I was on my way, bounding over the corrugations at the best speed poor Laura could produce, and was soon pounding on the door of the curator's sanctum. Dr. Leakey himself was away on safari as was John Williams, the curator of birds, but a girl assistant showed me into a small office where a tea chest, the open end secured by a metal grille, was standing in the manner common to beast-containing tea chests the world over. On my knees in an instant I peered expectantly into the dark interior, to be greeted by a pair of huge lustrous eyes and a small delicately made hooked beak, now open in hissing defiance. It was like finding Cressida all over again, except that the as-yet-unidentified occupant of the chest was certainly no kestrel. I thrust forward my gloved left fist and, as I had expected, the hawk threw itself on its back and clutched the glove with two sets of sharp curved talons. With my right hand I got a comfortable hold of the wings, tail and body of the bird and eased it carefully forward and upward, thus disengaging the talons which struck out wildly, seeking a new hold as I drew their owner out of the dim recesses of its prison.

I was amazed as I looked at the little hawk. Turning her

head to glare fixedly straight into my eyes, she was obviously both amazed and displeased. Never before had I seen a bird like this. In shape and general proportions it resembled a hobby, but was just that little bit larger. She was the color of a passage peregrine, but cinnamon-tinted on the breast and with conspicuous white cheek patches, which precluded her from being an African peregrine. Besides, the shape was different, lighter and more delicately made; the wing tips reaching beyond the slender kestrel-like tail. Examination of the mounted specimens on display and the skins in the drawers soon proved that she was neither an African hobby nor a red-legged falcon, and she was certainly not a lannerette. Furthermore, she did not appear to belong to the numerous subspecies of the peregrine falcon; the long narrow tail and the generally darker coloring was enough to show that she could not be that beautiful chunky little hawk. For the moment I gave up; it was enough that once more I had a falcon under my care.

My chief concern was to get her back to camp and to restore her to full flying condition. Apparently she had been picked up on the side of the road where she had collided with the ever-present, ever-destructive telegraph wires just outside the Nairobi suburb of Karen. I needed a name for my new acquisition; what could be more appropriate and attractive than Karen? Like Cressida before her she was emaciated and ravenously hungry. I bought a bit of best beef at the first butcher I came across and there and then, sitting on the passenger seat of the land rover, she tucked in with a will and ate until she could hardly stand. This method of providing unlimited food for a semistarved bird may not have been in the best medical tradition, but it was uncommonly satisfying both to me and to Karen and it certainly had no ill effects. She was placed carefully in a sock-lined cardboard box and I hurried happily back to Kajiado.

As I loped toward my tent carrying the precious cardboard

A Home for Karen

box, Sally, who had been standing look-out in the upper branches of the umbrella tree, came trundling up, twittering with pleasure and affection. She sat on her hunkers beside me on the bed as I untied the string that kept the box hawk-proof. I picked up Karen and put her on my knee, covered with a towel. Sally put out a gentle inquisitive paw to touch the new arrival. Like a miniature tigress Karen threw herself on her back and struck out, chattering with fury. For a second the talons of one foot closed on Sally's intruding hand. Sally let out a shriek of fear and indignation and fled through the tent flap, knocking over table, chair and first-aid kit as she went. Cursing her roundly I collected my bandages, scissors, elastoplast and bottle of TCP and got to work.

As far as I could make out the only trouble Karen suffered was severe bruising of the carpal wing. It certainly did not appear to be broken and she could scuttle about on the tent floor using both wings to hasten and assist her progress. The wing mechanism, however, is a delicate and complex organism and I was going to treat it as more serious than might in fact be the case. Obviously the more rest the wing had, the greater the chance of recovery. In order to render the injured wing immobile I put both wing tips together in a completely natural position and fastened them with a bit of sticking plaster. This was by no means easy as I had no one to help me and enjoyed only the moral support of Sally, who had by now returned and sat watching the proceedings with what seemed to be the deepest sympathy in her expressive hazel eyes.

Baboons, being sociable animals and very sensible too, are devoted to each other, and will if necessary risk their very lives to help one of the troop who may be injured or wounded. They will also unite to rescue one of their number who has been captured by their arch-enemy, the leopard. This approach is entirely different from that of other gregarious beasts, such as antelopes and wolves, which will either ignore or attack and drive off any unfortunate who happens to be incapacitated or abnormal in some way. As far as Sally was concerned, Karen belonged to me and was therefore part of the family. As such she was to be treated with respect and good humor even if Karen herself had other ideas!

That night I lay awake trying to remember all I had read about falcons and falconry. The only book on the subject I then owned included neither pictures nor descriptions of any bird resembling Karen. I seemed to recall a book I had owned in my boyhood called *Lord Lilford on Birds;* it was all about the 19th-century sportsman and ornithologist who, among other accomplishments, had been a keen and highly proficient falconer albeit confined to a wheelchair for the last years of his life. He described a falcon he had once owned that had brilliant white cheek patches and had been an outstanding performer. The name slowly formed in my semiconscious brain. "Ellen's falcon?" No. "Eleanor's falcon?" Wrong again. At last the penny dropped, "Eleanora's falcon," that was it. Surely Karen could not possibly belong to that species, which I believed was confined to a few islands in the Mediterranean. Still wondering, I fell asleep.

Early next morning Karen was very much alive and, whatever her species, active and uncommonly hungry. I lifted her onto my left fist and gave her a substantial lump of beef. After looking at me with lofty condescension, daring me to make a false move, she began to feed, gripping the glove with her powerful feet which I noticed were remarkably large for her size. She herself was a gorgeous creature, beautifully streamlined, and she gave the impression that, once restored to full health, she would be capable of devastating speed and aerial dexterity. I had watched hobbies in action both in Europe and in their winter home in Africa and knew what they could do when so inclined. This bird looked as if she could surpass the performance of any hobby that ever flew, with real clutching falcon feet, quite different to the short stumpy mouse-catching toes of a kestrel.

She was also ridiculously tame and self-possessed. After eating until she couldn't swallow another sliver of meat she looked me straight in the eyes with supercilious arrogance, roused her feathers, let them slowly settle back into place and rubbed her beak confidentially on my thumb; she then tucked one foot into her breast feathers and turned her head to examine the sticking plaster which held her wings together. I made and put on a pair of light jesses and settled her on a low block in a corner of the tent; only time and rest would tell if Karen would make a complete recovery and fly normally again.

She appeared to be a mature bird, a haggard, and I could not help conjuring up pictures of some pine-scented, swallowtail-haunted headland high above the azure, white-flecked Mediterranean, and of a falcon, an arrowy shape, black against the glimmering sea, swinging in low and fast against the gently undulating waves to rocket up and disappear into an inaccessible cranny of the granite face above. All this was pure speculation, of course, for I had only the haziest idea of what sort of raptor Karen might be, but it was singularly satisfactory rumination nonetheless. I decided at least a fortnight must pass before I removed the restraining plaster because, should the bone have been fractured, the slightest jolt or displacement would ruin the whole operation. Meanwhile she sat on the block, ate like an eagle, and became apparently content.

I made my report to the Locust Control top brass in Nairobi and they at least appeared to be satisfactorily impressed with my account of the trail-blazing in which Zacharia and Laura had acquitted themselves so well. However the idea of making a road was given up, and no doubt the whole area is as it was then—an arid, untrammeled wilderness.

About this time a new menace, as destructive but more

insidious than locusts, was beginning to make itself felt. For some years the name Mau Mau had cropped up from time to time but no one seemed to know much about it or what it meant. The settlers and officials were well aware that something was going on but just what it was and how it was to affect their lives and indeed the whole future of Kenya and its people, both European and African, was yet to be made manifest. The very nature of the insurrection and of those involved in it was, in those early days, far from clear. There were isolated instances of violence; a herd of cattle hamstrung, another brutally cut about with *pangas*, a lonely farm attacked and the occupants butchered. But none of these happenings seemed to be connected and the driving force behind them was obscure. What was however abundantly clear was that something evil was sweeping through the peaceful and prosperous land of Kenya.

For a while this did not affect me. I was in the Masai country and the Masai were the traditional enemies of the Kikuyu (who were suspected of being at the root of the trouble). In any case the Masai were a race apart, proud, disdainful and far too disinterested to become involved in any sort of subterfuge. If a Masai warrior didn't like you he would tell you so, and if you really annoyed him he would stick a spear through you, but he did it openly and from the front. He would never sneak up from behind and knife you in the back; these things were left to lesser men. Meanwhile life went on for me much as it had for the past two months.

Exactly two weeks after I had taken charge of Karen I removed the plaster from her wings; I snicked through the middle where the wing points were held confined, and settled her on my bed while I fetched some hot water to soak off the last traces of her bondage. When I returned with the water she was preening unconcernedly and the wing had flipped back into its natural position. I put her carefully on the back of

A Home for Karen

my solitary chair and watched as she mantled, stretching first one wing along the leg on one side, and fanning her tail to its full width while standing on the other leg; and then repeating the whole performance with the opposite two members. It was a particularly satisfying piece of self-indulgence to watch, giving me much the same feeling as I get when I see a well-fed cat, curled up asleep in the sunlight.

I attached a light leash made from a football bootlace to her jesses and carried her, chair and all, into the sunny world outside the tent. Soon she was exercising her wings happily, until she was dancing as lightly as a ballet dancer at the full stretch of her black shiny talons and yellow toes. Dingaan, who must have been watching this performance from afar, arrived and swore at her, using the foulest shrike language he could lay his tongue to; he dropped to the ground and did a war dance all around the chair while Karen, her feathers tightened close to her body and a fierce look of anticipation in her fiery eyes, leaned slightly forward, her head bobbing with somewhat sinister intent. I shall have to keep an eye on these two, I thought. How right I was!

Training Karen was simplicity itself; already manned by her weeks of inactivity and my constant presence she was quick on the uptake, and in no time at all would fly across my tent to the fist for rations, with a light line or creance to act as a drogue anchor. She was soon coming ten, twenty, fifty and finally a hundred and fifty yards across the open veldt. Now I introduced her to the lure, in this case the dismembered wings of a laughing dove found conveniently by the side of the road on the way to Mananga. These wings were attached to a contraption of cowhide to which short leather thongs had been affixed to hold the actual reward, a small piece of tender and prominently displayed beef. Karen would come like a bolt for this curious fabrication, dropping upon it as it lay meat-side up on the grass of Arthi plains. She was just as quick to follow the lure when swung for her above my head, but to keep it out of her clutches was another thing altogether. She was unbelievably fast and nimble and when flown free would almost turn herself inside out in her generally successful attempts to catch this rather unattractive replica of her legitimate quarry.

If there exists anywhere in the world a better place for practicing the art of falconry than the Masai plains of Kenya, then I would be delighted to see it. Mile follows mile of gently rolling grassland with here and there a distant outcrop of rock to break the monotony, while closer at hand lie a few isolated acacia trees with their promise of shade in which to recuperate after a hard-fought flight. Here were no prying eyes or inquisitive spectators, just the sky, the plains, the gentle sweet-scented African breeze, the far-distant horizon, and your falcon waiting on high overhead, a tiny ever-watchful black arrowhead riding the breeze and drifting in seemingly aimless circles as she waits for you, her friend and partner, to flush the quarry, dove, quail or spar fowl from the tangled, lion-tinted grass far beneath her.

I had not been flying Karen long before I noticed several unusual things about her. She was, as I had expected, unnaturally speedy in the air; she loved flying and would put up an amazing display of aerobatics apparently for pure devilment, burning up the sky like a large supercharged swift, but with an additional air of menace that no swift could ever hope to emulate. She was also, to my considerable surprise, largely nocturnal or at least crepuscular in habits, and would become increasingly active toward dusk and, rather more inconveniently, when the first light of dawn with its ghost-white mists heralded the coming of another glorious day.

This owl-like habit of Karen's was soon put to good effect. Among the alcoves and crevices of the district offices of Kajiado there dwelt a colony of epauletted fruit bats. These intriguing beasts were like miniature editions of the well-known and unpopular flying foxes of the Far East. They are, however, not much bigger than the European noctule, but have curious heads like rather bulbous-headed Alsatians and they get their name from the tufts of white hair which spring from their shoulders and appear to be purely ornamental. These bats were, as their name implies, fruit eaters, and as the nearest plantation of available fruit was miles across the open plain they used to issue forth well before dusk and go flopping off with their deliberate, unbatlike flight, traveling in columns like miniature bombers from some satanic nocturnal air force, with none of the erratic zig-zag irregularities that one associates with the small insect-eating bats that flicker so characteristically about a British summer sky at dusk.

I had often seen these bats or a similar species before and had speculated on what sort of show they would put up if chased by a hawk. I had vaguely imagined that bats were so adroit on the wing, aided by their peculiar system of built-in radar, that few if any birds could catch them. Then I remembered the Anderson's bat hawk. This unusual and highly specialized creature, which looks rather like an enormous nightjar but flies like a true falcon, is I believe the only truly nocturnal hawk. It is incredibly adept at catching bats, hence the name, and in fact feeds on little else. Well, if one bird could do it, surely another as well-proportioned and equally adroit could also do so?

It was easy to enter Karen to bat, the ones roosting under the D.C.'s office porch being a breeding colony. Like all bats they carried their youngsters about with them on their nocturnal forays until the little ones became too large for this kind of transport; but for some reason they were very careless, often dropping their babies twenty feet at the roost. I used to visit the porch early in the morning after the bats had returned, and I would be almost certain to find two or three well-grown youngsters lying cold and lifeless on the ground. This suited Karen, who appeared so eager for a diet of fresh bat meat that I wondered if perhaps these beasts were included, if only rarely, in her natural diet in the wild. I was determined to find out.

I knew exactly when the bats would emerge and the direction they would take, for I had been watching them on and off for weeks. I knew where their feeding ground was, and I had even seen them there, stuffing themselves with ripe guavas till they could hardly fly. One clear, calm evening I stationed myself out on the open plain with Karen on the fist, perhaps three quarters of a mile from the District Offices. There I awaited the arrival of the unsuspecting horde. Almost to the minute the first echelon arrived, traveling purposefully with slow flapping wings at deceptive speed. They were flying low, perhaps thirty feet above the grass, when I cast Karen off. Up she went, climbing in wide circles, her wings flickering as she strove to reach the upper air. The higher she rose the smaller became the circles she was making; but she was also traveling in the opposite direction to the bats and there was no indication that she acknowledged their existence. I was getting somewhat anxious and was about to produce the lure. Darkness in Africa comes down with disconcerting speed and I did not want the rapidly climbing Karen to be swallowed up in the gathering gloom, for I doubted if I would ever catch up with her again now that she had reached her pitch and was resting on her wings away up there in the deep purple twilight, as self-assured as a bat herself.

Without any warning of her intention she swung round, paused for a second to sum up the opposition, and hurtled out of the sky, dropping like a well-aimed boomerang upon the rapidly retreating ranks of her quarry, now almost out of sight. As she disappeared behind a fold in the ground, all of six hundred yards from where she had begun her stoop, I pounded after her, falling over the jagged rocks, only able to estimate her probable position. I could see no trace of her in the sky above, and now the last glimmer of light was about to depart. I had had the foresight to bring a torch and I had a white handkerchief in my pocket. I fastened the handkerchief to a stick and forced the end of it into the hard unyielding ground at the point where I assumed she was most likely to be. She could of course have gone raking off into the night—this was more than a possibility.

Using the handkerchief as a marker I moved out a hundred yards or so and began to circle the stick slowly, working my way inward and examining as much of the ground as the thin white beam of the torch would allow. As I walked I almost regretted that Karen was not wearing a bell on her leg as all the best falcons are supposed to do. Personally I never have and never will use them myself, because I am convinced that they are just one more hazard to overcome should a hawk become lost and have once more to start earning its living in the wild. I kept on searching, pausing to listen for the slightest rustle among the wiry grasses. It was only by chance that I blundered onto her, where she crouched in almost total darkness still mantling her quarry in the open mouth of an antbear hole; only the vivid white of her cheek patches betrayed her presence. I lifted her, prey and all, onto my fist and hurried back to camp where she continued her interrupted meal, undismayed by the glare and hiss of the pressure lamp or the hurrying white-robed figure of Zacharia as he brought my evening meal.

A few evenings later, feeling a bit off color from what I suspected to be the onset of an attack of malaria, I took Karen out onto the plain for some exercise. I had recently run out of Mephocrine and Kajiado was supposed to be more or less mosquito-free, but I knew by now the unpleasant symptoms, the slight but persistent shivers, and the sudden wave of nausea that heralded a new onslaught of the beastly bug. Anyway Karen had to be flown and I decided a few minutes hard practice at the lure would be better than nothing. I cast her off and she climbed at once, putting a couple of girdles around the camp in her characteristically sizzling style. A lone bat appeared from the direction of the District Offices and Karen stopped at once. As the bat swerved from her attack I caught a flash of white wings. I had seen bats like this in Lamu; it was one of the silvery-gray white-winged insect eaters, a species with infinitely greater speed and maneuverability. Karen accepted the challenge and the chase was on. As the bat jinked from the first stoop Karen threw up vertically many feet, turned and stooped again with the zip of a rifle bullet. The bat dropped almost to ground level, zig-zagging in a way no bird could hope to copy. Momentarily at a loss, Karen mounted once more and followed, parallel with and about twenty feet above her intended quarry, who was now just skimming the grass and virtually unassailable.

It was now apparent that the bat intended to return to its roost, no doubt to wait until the night air should be free from such aggressors. Evidently thinking that Karen had given up the chase he rose well clear of the ground and made a beeline for sanctuary only half a mile or so away. Karen must have been waiting for this moment and made a further attack, but the bat's special radar system warned him in time for it looped the loop and Karen missed once more. The bat was almost as fast and a great deal better at swerving and diving than Karen, who was by now intent on finishing the affair as quickly as possible. This maneuver had given the bat enough time to pull away from Karen, who was by no means restored to full fitness yet after her enforced idleness and was panting with the effort, her beak wide open as she made a final

A Home for Karen

effort to fly down her quarry in level flight. I had forgotten my malaise as I watched the two well-matched opponents fight it out over the rapidly darkening plain. It was like an airborne coursing match as the bat, his wings a creamy blur, made every effort and used every trick he knew to gain safety. Karen, tired as she was, traveled across that plateau at apparently undiminished speed and drive, but the bat won in the end, disappearing among the white arabesque buildings that I could still just make out behind a screen of tall slender acacias. Karen returned but doubtless not before trying to follow her fleeing quarry into some inaccessible cranny. She came down to the now almost invisible lure, and I was interested to note the ease with which she could pick it out as it lay half hidden in the grass. Remarkably complacent, she did not seem put out by her gallant failure, but ate an enormous meal by way of consolation while I went to bed to await the expected onslaught of fever.

On emerging a few days later, weak, wobbly-kneed, washed-out but alive, I decided to accept the long-standing invitation of a friend who lived in the Nanyuki area beneath the shadow of Mount Kenya, to spend a few days recuperating and, at the same time, hobnobbing with my opposite number in that part of the world. Zacharia would be close to his own people once more; he was none too happy among the Masai whom he feared and distrusted. Both Zacharia and Sally had proved themselves during my illness. Zacharia had cooked, washed, fetched and prepared medicine, kept me supplied with much-needed cold drinks and had done all and more than could have been expected of him. Sally had hardly left my bedside except to answer the calls of nature and had proved a warm, loving and sympathetic comrade. I would awake from a nightmare-ridden delirium to find her woolly gray body and warm clasping hands cuddled close beside me on the narrow bed, her hazel eyes searching mine with pathetic anxiety. Her delight on my recovery was heart-stirring. Dingaan was glad to see me about again, but wondered what all the fuss was about. As for Karen, she was as aloof and dispassionate as ever; she was a true falcon and therefore an aristocrat through and through, and seldom gave one an inkling as to what her true feelings might be.

11

Mau Mau Menace

NANYUKI, forested, hilly and clear-aired was quite different from Kajiado; it lay on the equator somewhere between eight- and nine-thousand feet above sea level whereas Kajiado was only half that height. The nights were therapeutically chilly, and blazing log fires the accepted thing. The rain forests were full of green-bodied red-winged maracos, with their clownish faces and their weird honking crowing calls, and with Colobus monkeys, whose handsome black-and-white coats, long thick tails like furry bell ropes and gentle almost other-wordly behavior make them the most glamorous and attractive of all Africa's primates. These forests were also the haunt of the rare and elusive Ayre's hawk eagle, the smallest but certainly the most dashing and courageous of all the numerous eagles that are still to be found in that area. I was lent a small log cabin of the sort one associates with the Canadian backwoods rather than the forests of Africa, but which suited to perfection the hot, brilliantly sunny days and jewel-clear, bitingly crisp nights.

Micky Williamson weighed about two hundred and twenty-five pounds and had been an amateur boxer of some repute. He was credited with having killed a leopard that had attacked one of his dogs, using nothing but his bare hands; had I been the leopard I would have gone quietly before the battle began. Micky, like so many of his type, seldom talked about his exploits and had only three passionate interests, his family, his farm and his Rhodesian ridgeback dogs of which he had a superlative array. The Williamsons were a hardworking family and I saw little of them during the day, but we used to foregather at dusk. In all probability a neighbor or two would drop in and we would drink a sociable glass of brandy and chat about the world situation in general, which in this case meant the Mau Mau.

My cabin stood in a small isolated clearing a good quarter of a mile from the farm itself. It was an ideal spot from which to observe and study the habits of the shy, elusive denizens of the forest. All sorts of game abounded there; the bush buck, leopard and serval cat hunted or were hunted among the great creeper-festooned forest trees, and even the rare bongo antelope with its spiraling, ivory-tipped horns was believed to haunt the remote interior, though few had ever seen it.

In a corner of the clearing in which the hut stood was a small overgrown garden, the main feature of which was a thriving intertwined mass of grenadillas; these had fleshy leaves and delicious fruit which grew in the form of Chinese lanterns. Late one night I was returning from a final prowl, and decided to take a few of the ripe fruits to share with Sally who was crazy about anything of that sort. As I reached out to pick the fruit I heard a sudden dry rustling from close at hand and just ahead of me; I had never heard this sound before and it startled me considerably. Then, from left and right it came again, loudly, almost menacingly. I groped for the torch I carried in my trouser pocket, hastily pulled it out and pressed the switch, almost fearful of what I might see. To my intense relief the sudden shaft of light revealed that which I should

have guessed: I was in the middle of a small sounder of porcupines. The huge sow, who appeared nearly as big as a badger, had all her formidable array of spines, as sharp as Masai spears, and was gazing in my direction; she grunted, stamped her forefeet and came trundling toward me, grinding her huge rodent teeth in a far from welcoming way. I glanced from left to right and saw two or three miniature replicas of herself, all holding tendrils of grenadilla in their mouths and all with spines raised truculently. Having, as a small boy, read Percy Fitzpatrick's *Jock of the Bushveldt* I knew what a porcupine could do when it didn't like you, and these certainly did not care for me. I beat it at the double, not caring to be impaled upon such a formidable armory of quills. I had once examined the body of a leopard, a notorious killer of goats, and realized why the poor brute had been forced to prey on such humble quarry. The whole of one forepaw was skewered through and through with these fiendish black and white spines, the barbed ends of which had worked deeper and deeper into the flesh until the unfortunate leopard must have been driven almost to distraction with the pain. I was glad that its sufferings had at last come to an end.

In all the years I spent in Africa, during which I must have wandered for many miles at night, often wearing nothing on my feet but rubber-soled gym shoes, I saw only a handful of poisonous snakes; these were mostly the hideous evil-looking but normally slow-moving and retiring puff adders which, unless one happened to step on one, are nonaggressive. Apart from the one incident when I nearly walked into a black mamba at Oldeani, as previously related, I had no narrow escapes. But my father, when in Somaliland, was once bitten on the ankle by a puff adder, and had just been able to rush into his quarters and grab a vial of anti-snakebite serum. His hand,

Mau Mau Menace

contracting under a spasm, broke the bottle before he fainted and the serum entered his bloodstream through the cuts, thus saving his life. An extraordinary story but perfectly true, so my mother assured me; and who am I to doubt her word?

I had to attend a conference in Nairobi on the breeding cycle of the desert locust and I would be away one or possibly two nights. As neither the lordly Norfolk Hotel, nor the almost equally plutocratic New Stanley, were likely to welcome my jolly menage, and as I could hardly expect my hosts to keep an eye on them during my absence, I arranged to stay with another family on the shores of Lake Naivasha, no great distance from the conference.

In the afternoon we went out in a boat to watch wading birds and marsh harriers. I have rarely seen so many waders, most of which I am ashamed to admit looked almost identical to me. On the next afternoon I flew Karen, to the astonishment of my hosts who, like so many spectators before and since, seemed flabbergasted that a bird, especially a hawk, should prefer to return voluntarily from its lofty station in the sky to the fist of its friend and trainer, rather than to shoot off to the furthest horizon at the first opportunity.

The conference ended and I said goodbye to my old acquaintances, the wild men from the far flung outposts of the Northern Frontier Division, from Lake Rudolph and from Kismu, hundreds of miles to the west on the banks of Lake Victoria. As the land-rover nosed her way eastward once more I had a curious premonition that grew stronger as the miles rolled behind us. I was glad that I had brought Zacharia, who had spent the time with his family, and was also relieved that I had brought Dingaan, Karen and Sally as well, even though the latter had disgraced me by slipping into the *posho* store of my friend's farm, breaking open a bag of *posho*, and covering the floor, the ground outside for yards around, and herself with pounds and pounds of the penetrating yellowish-white meal.

I drove down a narrow tree-lined track that led to the cabin, and as we rounded the final bend a savage acrid smell bit into my nostrils, a sudden surge of apprehension sharp as a dagger swept through me. The headlights picked out all that was left of the hut. A jagged rectangle, perhaps three feet high, black and smoking, was all that remained. The ground, for yards around, was aglow with the still burning ash. I looked at Zacharia; his face in the dim light was gray. *Kitu m'baia kabisa. Shauri ya Mau Mau* (this is a shocking thing, the work of the Mau Mau) was all he said. Together we cleared a path through the red-hot ash, but I knew well enough what to expect.

I was right; nothing was left. Of my few personal possessions only the twisted clasp of my trunk remained. I had not owned much, for I have a rule to travel as lightly as possible. My clothes were charred beyond recognition, but what really sickened me was that all my photographs of Cressida and Bracken had been destroyed and with them the much-treasured medallion that I had received from Dorothea St. Hillbourne, then head of the People's Dispensary for Sick Animals, to mark Cressida's inauguration into the Allied Forces Mascot Club shortly after her return from Germany. I think this must have been stolen along with my binoculars. The only photograph that survived was one taken in Arusha of Rupert and Torquil because I had kept it in my wallet in the pocket of my bush jacket. Battered and creased as it is, I still possess it; my only link with the Africa I knew and loved.

Micky Williamson and his family had just returned from a settlers' meeting when they heard shouting from their farm hands and saw the ominous red glow in the sky. Micky had grabbed a gun, called his dogs and gone in pursuit, but it had been the work of moments. The raiders had struck the wood cabin, inflammable as a tinder box, done their work, and melted away into the impartial shelter of the forest. Pursuit was hopeless; besides, the damage had been done. I felt a great surge of relief that I had at least taken all my animals with me.

The following day I drove to Nairobi, handed in my notice and applied to join the Kenya Police. I was goaded into joining the police because of the outrage perpetrated on my property and a very natural desire to do something about it. In an interview I was suddenly asked if I would mind seeing someone being hanged. I confessed that the prospect was not one that filled me with enthusiasm, but they did not seem to hold that against me, and I was accepted. I went back to Kajiado to work out my notice.

I also had to do a bit of organizing. This was going to be a very different kind of job than anything I had tried before, difficult and quite possibly unpleasant. For one thing I could hardly take up my new appointment accompanied by a miniature Noah's Ark. A dog would have been all right, possibly a hawk and even a shrike; these could at least be kept out of sight and in the background during the preliminary stages. But what of Sally? Besides, I had little idea as to what the work would involve other than that it would be highly mobile and perhaps dangerous as well. I would have to hand over my land-rover on the termination of my work with the Locust Control, but I still had a few more days to go and I meant to make the most of her while I had the opportunity.

I thought of Van Royen and the rumbustious extrovert Rajiki, and so I drove over to Oldeani with Sally. Van Royen was delighted to take care of her, and Rajiki took charge completely. Sally followed where he led. As I said goodbye to her, she put her arms around my neck, nibbled the lobe of my ear, smacked her lips and looked into my eyes. I tried to explain the position and I think she accepted it in her own unexpectedly adaptable way. As I drove out of the yard I looked at Sally for the last time. She was sitting on a low wall, cuddled close to Rajiki. They had their arms about each other and he seemed to be telling her about the wonderful new life that lay ahead for them now that they had found each other. I kept in touch with Van Royen, and I had news of how Sally had settled down and of the various escapades that she and Rajiki got mixed up in. Some years later in London, I got a telegram that read, "Rajiki + Sally = Kijana." I often wondered what the Post Office people made of it!

I brought the land-rover into Nairobi and handed her over to the Locust Control Transport Officer. I shall always remember Laura, KBE25. Seldom can a so-called inanimate object have developed such personality, given such yeoman service, and inspired such affection in its driver and temporary owner. I often wondered what happened to her and what other adventures she had, but I doubt if anything she may have accomplished in the future could have equaled that trip across country from Voi to Amboseli. I said a sad goodbye to Zacharia who was staying on with the Locust Control. He had wished to accompany me on my next employment, but as an African civilian this was not considered possible. He had proved himself an exceptionally loyal and versatile fellow.

Shortly after I was told to report for duty to the Internment Camp at Manyani, smack in the middle of the Tsavo National Park and not far to the east of Voi, which I knew very well indeed. I was issued a brand new uniform and was now masquerading as, I suppose, the equivalent of a second lieutenant, with a solitary metal bar on each shoulder and a lethal-looking .38 revolver in a smart leather holster swinging dramatically on my hip. Thinking I might as well do the thing in style I acquired the few medal ribbons to which I was entitled and stitched them onto my light khaki tunic. Thus arrayed in martial splendor I caught the night train from Nairobi and went creaking, jolting and puffing my slow way to a new career.

I wondered what I had let myself in for this time. On the seat opposite me Karen and Dingaan rested in two cardboard boxes. Although their possession might well complicate matters I nonetheless drew comfort from their presence. I had, many years earlier, learned to accept the fact that I was apt to be regarded as a harmless eccentric at the best. After all, I had my life to lead and I intended to extract as much enjoyment from it as I could. For this I had fought and had spent more than two years languishing in an assortment of prison camps, and nobody was going to tell me how to run my life now. I remembered too that whatever the initial suspicion with which I and my retinue were frequently regarded, this usually gave place sooner or later to warm friendship and an interest second only to my own.

Manyani was the site of the largest detention camp in Kenya; here were concentrated some thousands of hard core Mau Mau, proved and suspect. I was now bound by the Official Secrets Act, and not wishing to sample the hospitality provided by the governor of the Tower of London, I feel it incumbent upon myself to treat the ensuing experiences with restraint. Suffice it to say that the establishment in which I found myself was known to its staff as Dodge City and certainly most of my colleagues appeared to be dodging something, maybe just their own pasts. It was a strange, unreal existence; alcohol was both cheap and available in unlimited quantities; everyone carried revolvers, and the resulting atmosphere was a cross between a bad Western film and a worse gangster movie. My job and that of my colleagues, members either of the Kenya Police or of the Prison Department, was twofold: to keep the detainees within bounds and equally important to fight off any attempts at release by armed bands from outside.

The camp itself looked exactly like the sort I had myself experienced in Germany and more particularly in Italy, where the summer temperature was something approaching that of the inferno in which I now found myself. The only difference was, of course, that now I was the chap with the gun, responsible for ensuring that the same number of men were locked up in their huts at night as had emerged from them in the morning. It was, to put it mildly, an uninspiring job. I had often wondered what my guards had thought of, and now I had a pretty shrewd idea—how to get away on leave as soon and as often as possible.

I made two particular friends there, one being the inevitable Scotsman, who I have found always turns up in such situations. Jock came from Paisley and was so Scotch that I could understand with difficulty about one word in every ten he uttered; still we became close friends. The third member of our small but select group was the almost equally ubiquitous Welshman, Taffy Evans, an ex-British copper from Swansea. About two hundred and twenty-five pounds of solid muscle, he was a useful man with his fists and a more than useful rugger forward who, like so many of his countrymen, talked, thought and no doubt dreamed about the game he loved so well. Jock, on the other hand, was an ice hockey fanatic and had, if I remember rightly, been a member of the Paisley Pirates. With his speed, dash and dexterity he must have helped that redoubtable team to many a victory. He likewise was by no means averse to a scrap although I must doubt whether the late Marquis of Queensberry would have approved of his methods when he got warmed up. How a peaceful, retiring and law-abiding fellow like myself ever came to be mixed up with such a pair of desperadoes I cannot remember, but I was to be more than grateful for their friendship. Funnily enough, although Jock and Taffy were city-bred with little experience of country life, they both became fervent falconry enthusiasts.

I have found this often to be the case. The townsman frequently proves to have a deep-seated interest in country pursuits and when given the opportunity will as often as not beat the countryman at his own game.

As each blazing, blistering and boring day passed I was increasingly thankful that I had brought Karen with me. It became a sort of ritual to take her out of camp in the evening and to climb up to a natural shelf, formed out of the hillside a few hundred yards above the camp. Here I would cast her off. Helped by the hot air currents, she would rise rapidly until she had reached her chosen pitch, whereupon she would range far and wide, high above the thorny scrubland, till something of interest caught and held her roaming vision, perhaps a flock of gray ring-necked doves returning from a distant water hole; or maybe a pair of handsome harlequin quail, moving beetle-like across the space between one patch of withered vegetation and the next, would attract her attention as she rode the breeze far overhead. Sometimes she would keel over in an amazing stoop which, however, she seldom pressed home unless exceptionally sharp set. Mostly she would fly just for the sheer pleasure of it, and to watch her was to share the joy we knew she must be feeling up there, untrammeled and unrestrained.

A ledge of rock in the vicinity was the nesting site of a pair of Abyssinian lanners, and one evening the male, the dapper, piratical, pink and gray lanneret, sallied forth to challenge Karen to an aerial duel, a challenge she was only too eager to accept. He was slightly larger, but she was so fast and fiery as to make him almost clumsy by comparison. Not that he was a sluggard in the air; he could stoop and throw up like a peregrine. He was heavier and more powerfully footed too; had he struck her a serious blow I am certain that would have been that. However, the impression I had as we watched this display of airborne athletics was that neither antagonist was anxious to bring the affair to a decisive conclusion, and neither appeared too keen to test the physical courage and prowess of the other. They weren't exactly playing but they were not taking things too seriously either.

Not, that is, until the arrival of the female lanneret. Drawn perhaps by curiosity and maybe wondering what had delayed her mate's homecoming, she appeared on the scene like an avenging angel; grim, silent and humorless, she meant business all right. Her mate, sensing her mood, dropped all pretense of play and the two drove in hard at Karen, stooping at her from every angle. Outnumbered and taken by surprise she was still the fastest and most agile of the three, but she was beginning to tire and lose the edge of her speed. Deciding to end the affair before there was a tragedy, we jumped up shouting and waving our caps in the air; the cast of lanners sheered off and, after circling as if half-intending to return to the fray, made for the precipice behind us. Karen, panting and heaving, came down into a low tree and thence, after a brief rest, to the lure, to be carried home in triumph after a contest such as few can have had the thrill of watching.

A month after my arrival I had to return to Nairobi to attend a revolver-shooting course, during which I was introduced to a Mr. Justice Ruddock who was attending assizes in Nairobi. He was uncommonly well named (ruddock being a country name for a robin), being short and round, with a bright eye and ruddy complexion. Furthermore he had a robust sense of humor, was a keen and intrepid rider to hounds and an enthusiast about most field sports. However, what most interested me was that he kept and bred Salukis, which he had brought with him from the Middle East where he had previously been living.

I had always been interested in and admired these wonderful desert greyhounds. They really are the terrestrial equivalent of falcons, Sakers in fact, having been used by the Arabs throughout untold centuries for coursing hares and gazelles across the treeless, rocky and sandy spaces of the Middle East and North Africa. With their wonderful far-seeing dark eyes and gentle, almost wistful air of timeless good breeding they even look like the noble falcons whose way of life runs parallel to their own. One of the oldest of all breeds of domestic dog they have remained, fashioned as they are for speed, stamina and adaptability to desert conditions, virtually unchanged for thousands of years, one of the supreme examples of man's ability to produce perfection by careful breeding when the necessity arises. Without the Saluki, the falcon, the Arab horse and the dromedary, it is doubtful if the wandering desert tribes would have survived at all.

Mr. Justice Ruddock asked me to lunch and an excellent lunch it was, although as far as I was concerned it could have been bully beef hash. I had eyes only for the hounds, which were reclining all over the floor and on every chair and sofa. With their silky feathery ears, plumed tails, glowing lustrous eyes and slender deer-like muzzles they were an unforgettable sight. Silver sable, red gold, grizzled cream and rich warm mahogany, they brought the Egypt of the Pharoahs into that sunlit East African dining room. There were two puppies for sale, a seven-months-old black, red and gold dog, and a silver grizzled bitch of five months. After much mental strife I chose the dog, whose name was Talah el Fidani.

We had coffee in the garden and Talah pressed a long slender enquiring nose into my hand and sighed deeply. I think he sensed that, from now on, he was to become part of my life and wanted to find out just what sort of companion I was likely to prove to be. You do not own a Saluki—the Saluki owns you and it is well to realize this at the outset. This relationship can and should ripen to a close partnership, but you are lucky to be accepted as an equal. If you can train and understand a hawk you should get along well enough with a Saluki;

their outlook is much the same. However rich and aristocratic you may be the desert hound has the edge on you for breeding and will not tolerate injustice or neglect. Not that he will bite you; he will just ignore your very existence until such time as you have apologized abjectly, after which he will favor you once more. Talah had never been outside the Ruddock's rambling garden and had never worn a collar and lead. I had bought both in Nairobi just in case and these were now attached to my unsuspecting new comrade. At seven months Talah was a well-grown powerful puppy, standing over two feet at the shoulder (he was eventually to reach twenty-eight and a half inches, one of the largest of his breed). At that age he was highly strung and as nervous as a yearling thoroughbred, but there was something about his wide open laughing jaws and merry eyes that bespoke a latent sense of humor and, as I was soon to discover, a pronounced will of his own which his gentle manners and almost feline grace belied. On the long grinding journey to Manyani he lay curled beside me on the seat with his beautiful gazelle-shaped head resting on my knee. I was proud and happy to have a dog again. He was in fact the first pure-bred I had owned. His Arabic name, though most impressive and befitting one of his lineage, was rapidly shortened to Tally, and Tally he was to remain for the next sixteen years.

Manyani must have suited him admirably; in fact it must have been close, both in physical characteristics and temperature, to the deserts of the Middle East, the cradle of his ancient race. On that journey we became friends and soon, as the days went by, that friendship ripened into love and close comradeship. I occupied a small cubicle-type room—my bed, wardrobe and wash basin being the only fixtures—and Karen sat on her block in the shade of the porch that ran the whole length of the European Officers Quarters. Dingaan, ever independent and self-contained, came and went as his mood dictated. On my return from Nairobi with Tally I introduced him to Karen (who had been well looked after by Jock—he having discovered unexpected finesse in dealing with such birds).

Karen was sitting unleashed on the back of my solitary chair, one foot hidden in her breast feathers, completely relaxed. Tally walked slowly around the room sniffing hopefully for signs of something to eat (he was a great trencherman), and Karen watched him with interest, her head on one side. After examining his new quarters and polishing off two tins of Irish stew he came over to me, and suddenly noticed Karen perched on her chair in the middle of the room. He came up leering and gently waving the snow-white plume of his tail; a long delicately veined muzzle reached up to inspect her, whereupon Karen leaned forward and tweaked it hard. Tally jumped back and looked up at me with a hurt, puzzled expression. I patted him and pulled his ears. He retired under the table and curled up on the pile of old blankets I had arranged for him. Soon the only sound in the room was his regular contented breathing and I knew that he was happy to be there with me. Within twenty-four hours hawk and hound had completely accepted each other and I could safely leave the pair of them alone together for hours at a time if necessary with no fear of finding signs of any tragedy on my return. It is strange, this natural affinity, but it must be deeply implanted, because for centuries Sakers, the true falcons of the desert, have been used by Arabs for hunting gazelles in partnership with these speedy and tireless hounds.

I have often thought that each member of the great greyhound family has its opposite number or complement among the diurnal or raptorial birds. Thus the Borzoi, the powerful but elegant hunting dog of Imperial Russia, with its vice-like jaws and allegedly uncertain temper, is surely the equivalent of the female golden eagle, both being trained to hunt the same quarry, the wolf. The Afghan hound is also aquiline in appearance and temperament and again hunts the eagle's natural quarry—hares, small antelope and foxes. What could be more like the smooth-coated true greyhound than the peregrine with its speed, well-bred lines, gentle but fiery nature and its determination in sticking to its chosen prey when once it has made its mind up? The intrepid lurcher, tough, versatile and cunning, with its willingness to take on anything at any time, is much like a well-manned goshawk, ideal hunting companion for the poacher or the squire. The small, fast, friendly whippet, with its affectionate ways and love of the chase, is roughly the counterpart of the merlin, even if the former normally hunts rabbits and the latter small birds. While to carry the simile further, even the charming, decorative but, from a sporting point of view, virtually useless kestrel, has much in common with the equally endearing and equally useless little Italian greyhound, the smallest and perhaps the most lovable of all the greyhounds.

From their first meeting Karen showed not the slightest fear of Tally, and he for his part evinced a respectful interest in her; he seemed to know that, like himself, she was a creature designed for and existing solely for the chase. When she was cast off he would do his best to track her, bounding onward beneath her flight path at speeds that must have at times exceeded thirty-five miles per hour. A Saluki at full stretch, feathered ears and plumed tail streaming in the breeze caused by its progress, and the effortless elastic undulations as it catapults forward mile after mile, must surely be one of the most glorious sights in the whole world of nature.

12

A Proud Saluki

TALLY was only seven months old when he arrived at Manyani. Within a few days of his appearance an incident occurred that nearly brought his life to a premature end. We used to cruise about the camp and the surrounding country in openbacked utility trucks that were ideal for all the purposes for which they had been designed, and a good few others besides. Very soon after he joined me, Tally developed a dangerous predilection for following these trucks in the same way that he would have followed a horse had he been living among the people of his homeland. He would lope along beside the vehicle at a fast canter, his head raised and his eyes alert for any sign of movement ahead. If he saw something that interested him he would lengthen his stride and go skimming over the sandy ground at a speed that soon left far behind the truck, which was restricted to twenty-five miles an hour in the precincts of the camp. He had a habit of unexpectedly shooting across one's bows in order to inspect the terrain on the farther side, a trick which in itself was likely to cause his early death, unless the driver was constantly on the watch. I tried to discourage him, but somehow he always seemed to escape from my quarters just when I was about to start a motorized patrol.

On this particular occasion I was returning from some job shortly after dawn, and Tally for once was standing proudly in the open back of the truck, eyes scouring the horizon for possible prey as I drove through the gates into the camp. Suddenly he leaped out of the vehicle, stumbling onto his nose, which ploughed a furrow in the sandy track; recovering in seconds, he was on his feet and away at full gallop after a gray slinking shape that I could just make out moving at a furtive trot toward some distant huts. I shouted and the jackal broke into a shambling canter as Tally, spurred on by my voice, really put the power on and swept down on this quarry which was by now well aware of its peril. A British red fox can travel when he wants to, and the African black-backed jackal is even faster, but even as a gangling three-quarter grown puppy Tally's speed was such that the gap between them was narrowing rapidly as he swooped hawk-like over that burned-up treeless terrain. However the jackal just made it to the first hut, and slipped under the raised floor as Tally came sliding and panting to a halt in a flurry of powdery sand.

Going flat out the truck was left well behind, and I arrived to find Tally bounding around the hut, thrusting his long nose underneath and giving tongue in the curious staccato Saluki bark that always seems to belong to a much smaller dog. The jackal refused to come out and fight so I grabbed Tally and heaved him, struggling furiously, into the cab of the truck, where he continued to shout noisy abuse. I lay on my tummy and peered under the hut; I could see the jackal quite clearly. He was crouching, watching me, and I noticed that he had saliva dripping from his jaws. Was it rabies? Or could it be due to thirst, caused by the heat, the excitement of the chase and the lack of natural water supply in the district? The jackal looked young, a mere cub, with its woolly puppy coat

and frightened eyes. I had a revolver with me, but I couldn't get into a position to shoot accurately and keep the target in view at the same time; in such a cramped position it was better not to fire at all.

The cistern that supplied the camp with fresh water was not far off and nearby was a battered enamel pie dish. I had always understood that a rabid animal was unable or unwilling to drink, due to some involuntary contraction of the throat muscles. I filled the dish with clear water and pushed it under the hut where the jackal still crouched, then I went to the mess to look for Jock or Taffy. Neither was there. Eventually I found Jock, holding a snap check of the detainees in a compound where there had been signs of insurrection. I told him about the jackal, and with hardly a word he fetched his pride and joy, a brand new .22 Hornet rifle. We returned to the hut where I had left the jackal; he had disappeared, so had the water. I have seldom been more relieved. Jock and I agreed to say nothing. The camp commandant, whose word was law, was not a dog lover and would undoubtedly have ordered Tally's destruction as a precaution. In fact, had Tally caught up with the jackal, I would have had to shoot him myself; the risk would have been far too great to have done otherwise. Later I confirmed that the chances of the animal suffering from rabies were extremely remote. I am glad to say that in all the years in which I lived in Africa, where the disease is endemic, this was the only time I became in any way connected with its terrifying manifestation.

There were other dogs in the camp, in particular I remember two Alsatians named Brutus and Deal. Brutus was a beautiful wolf sable and Deal an equally handsome black and gold. It is hard to say who hated each other with the greater intensity, the two dogs or their owners. The former fought savagely at the slightest excuse, while their masters hardly spoke to each other if it could be avoided; and each was heard to

A Proud Saluki

utter threats of direct vengeance should either of the dogs be injured by the other in one of their interminable battles for supremacy. Tally at this stage abided by the old army maxim that it is better to be a live coward than a dead hero and kept well out of the way of both, a characteristic I thoroughly encouraged. I remembered the remark made to me years before by a certain boatman in the Lake District, who had in his youth kept a notorious strain of fighting Staffordshire bull terriers, but had long since given up the game. "Well, sonny," he had said to me when the question of dog fights had arisen in casual conversation, "if you're going to keep a fighting dog you've got to be a fighting man."

One night, when the temperature was at its insufferable hottest and driest and tempers were even more edgy than usual, the climax came. Jock, Taffy and I were lounging on the verandah, a modicum of ice-cold Tusker beer beside us, when suddenly three revolver shots cut through the night. We rushed out to find pandemonium. Outside the mess, which also did duty as a bar, the owner of Deal was lying on the ground unconscious, while Brutus's owner, a huge Afrikaaner, loomed over him. Brutus, his side heaving alarmingly, lay stretched in a pool of blood oozing from three holes in his shoulder. He had been shot at point-blank range, the bullets having passed right through him, narrowly missing the usual evening gathering at the bar. Brutus's owner, helped by awed spectators, carried the apparently moribund dog to his car. Without a word he jumped into the driver's seat and roared through the night to Nairobi where the nearest vet had his practice. Amazingly, the dog survived and made a further brief appearance at the camp a few weeks later. Deal's owner, with a broken jaw and dislocated shoulder, went to the hospital in Mombasa and was seen no more at Manyani. Deal himself was adopted by another police officer and went to live with his family on the coast at Nyali beach, close to Mombasa.

A few nights after this unpleasant incident a meeting was called in the mess, to be attended by all the European officers who were not on guard at the time. The commandant, who held a rank roughly equivalent to that of a lieutenant-colonel, rose to his feet and spoke long and loud. He was not, he declared, going to have his officers behaving like savages. Fair enough. However he went on to say that, in future, no dogs would be allowed to roam loose about the camp and that any found unattended would be shot on sight. The only dog in camp at that time, other than a few pi dogs owned by African askaris, happened to be Tally, so presumably the edict referred to him. I rose to my feet, followed at once by Jock and Taffy; but before I found words suitable to the occasion Jock, who was quick-tempered and inclined to be aggressive even when sober (which he was not by any means then), turned to the C.O., who was seated at the far end of the table flanked by his myrmidons.

"I hope, sir," he remarked quietly, "that you are not referring to our dog, Tally."

The C.O. looked at him with amazement. "I am referring to any dog found loose in the compound," he replied.

"Well, sir, I would not advise you to shoot our Saluki. I'm telling you, that's all!"

Then he sat down, amid an echoing silence that seemed to rebound from the very walls. I was so stunned by this turn of events that I could only mutter something like, "That goes for me too!" Taffy for once said nothing, but looked uncommonly menacing, and the meeting broke up amid general clamor.

Within a week of these happenings Jock and Taffy were both posted away from Manyani, each being sent to the Kenyan equivalent of outer Mongolia, the Northern Frontier Division.

A Proud Saluki

As for myself, the following morning I handed in my notice, something I had been intending to do for some time. I felt horribly guilty about Jock and Taffy, but both told me that they could not have stood Manyani much longer, and that they were going to apply to be moved elsewhere in any case. I never saw either of them again, although for some years afterward I would receive letters at irregular intervals telling me the latest news of their whereabouts. Both finished their contracts and I believe both joined the Malayan constabulary. No wonder the crisis in that unhappy peninsula came to an end shortly afterward. It would have taken an uncommonly tough band of Communist terrorists to have tackled those two.

I now had two months to fill in and it was by no means too long for all I had to do. I wrote to the Canine Defence League, the R.S.P.C.A. and any other association I could think of, asking them to arrange accommodation for Tally in the best quarantine kennels in Britain. Eventually I managed to get a vacancy for him in the Blue Cross Kennels at Shooters Hill, near Woolwich.

The biggest problem was how to get him home in time, as I was anxious for him to complete a good slice of his enforced imprisonment before I myself arrived. I wrote to every airline that included Kenya in its itinerary and they sent me specifications as to the size of the crate I would have to construct for his journey. In those days the regular jet flights about the world, which one now takes for granted, were only just beginning, and the better-known lines appeared to be none too keen on carrying livestock. I soon discovered that it took about three days for the trip, and the idea of Tally being confined in a crate, however large, for three days was unthinkable. So I decided the trip would have to be a sea voyage after all. I had met other dogs that had traveled long distances on big ocean-going liners and they seemed to have had a pretty good time, though admittedly in these cases their owners were traveling on the same ship. I had to chance it. Tally was still only about nine months old and, with a life expectancy of at least twelve years ahead of him, I considered it a risk well worth taking. I finally arranged a passage for him on the S.S. *Kenya Castle*, sailing in ten days time, though I hated the prospect of leaving him in strange hands, however kind and sympathetic.

About this time a young man named Grant arrived in the camp. He was Kenya born and bred; in fact his father, a D.C. at Kajiado shortly before my arrival there, had been killed with a spear by a young Masai Moran, who disagreed with his method of meting out justice. The dispute had arisen over a certain white heifer, which belonged to the Moran and which was included in a forfeit that he had to pay for some misdemeanor. The Moran was particularly fond of this heifer and asked the D.C. to take some other animal instead. The D.C. was adamant, and the Moran, in a sudden rage, impaled him with his terrible fighting blade. Young Grant, however, did not seem to bear any grudge. He was at this time a member of a carefully selected and highly skilled group, whose job was to interrogate the detainees and to sort out the "sheep" from the self-confessed "goats," who had taken all the Mau Mau oaths. He must have been ideal for the job because not only did he have a deep and abiding love of Kenya and everything concerned with that wonderful country, but he also spoke Kikuyu like a member of the tribe.

Grant (I cannot remember his Christian name) and I found much in common and, missing the company of Taffy and Jock, I spent a good deal of my off-duty time with him. He was very interested in Karen, but like so many others, seemed to believe that falconry as a sport had more or less died out with the invention of the crossbow, and that anyone practicing it

A Proud Saluki

must have more than a touch of the weird and mystical about him. He had obviously never encountered the redoubtable Captain Knight or the equally practical and outspoken Gilbert Blane. Karen put on a good show for him and he was duly impressed. We could not, however, find a suitable quarry for Tally, all the game in Tsavo Park being strictly taboo.

The last week before Tally was due to sail from Mombasa we all motored up with Grant to the headquarters of his organization, which stood on the Arthi Plains between Nairobi and Kajiado. Both Karen and Dingaan traveled with us, of course. Dingaan was pleased to get back to the Highlands after months in the enervating heat of the Tsavo country. Here, once again, were many fiscal shrikes—at least one perched sentinel-like on every bush and telegraph pole. Although I had not realized it, Dingaan must have matured remarkably during our nomadic life together. I had never put any restraint upon his movements and he had always pleased himself, coming and going as he felt like it. Although still as tame as ever, always willing to come when called and ever ready to accept a tempting tidbit, he was by now completely independent and self-supporting, remaining with me only because he happened to like me and, no doubt, because our association had become a way of life for him. Now, back in his own natural haunts once more (fiscal shrikes are rare or absent from the low-lying coastal bush country) he felt an urge that had previously been lying dormant.

The very morning after our arrival he did his best to persuade an attractive young hen shrike that he was the handsomest bird that ever flew and, after a certain amount of initial bickering, he seemed to have succeeded. Returning that night to sleep on the open cupboard door, he left at first light and, flying to the top of a convenient thorn tree, serenaded his newfound mate with the clear, faultlessly beautiful shrike song, as he gave vent to all the emotion that neither he nor

I knew had lain so long concealed within him. I was both surprised and relieved at this outcome of our long and eventful relationship, but more especially I was happy for him, that now at last he had found his rightful niche. That his mate accepted him was certain because, throughout the remainder of my short stay, I saw them constantly together. Dingaan, on several occasions, purloined small pieces of liver intended for Karen and fed his mate, who fluttered her wings and opened her beak in the typical pleading gesture which seems to be the prerogative of fledglings and females among the passerine birds. Several members of the staff of this camp were keen ornithologists and promised to keep an eye on Dingaan and his mate, and to let me know what the eventual result of the union might be.

Grant and I got up early on the second morning and mounted a couple of tough Somali ponies. My steed was a beautiful gray, a frolicsome filly with the attractive, if somewhat incongruous, name of Silver Wedding. We called Tally, who fell in happily beside my pony, keeping about level with my left stirrup as we moved out onto the silent dawn-shrouded grassland which rolled away mile after mile in all directions. We moved first at a fast trot and then, after Silver Wedding had amused herself and relieved her feelings with a couple of light-hearted bucks that all but unhorsed me, we broke into the long, smooth, distance-devouring canter that is so characteristic of the Somali country-bred, and shared only by its near relative, the Arab.

Far ahead a small group of Thomson's gazelles, the ubiquitous Tommies, were feeding, tails a-flicker, in the curious restless way they have. Tally saw them and surged forward like a torpedo, after rearing twice on his powerful hind legs to get the distance. Grant, on his bay gelding, was now travel-

A Proud Saluki

ing at full gallop and I gave Silver Wedding her head, trusting to her inherent good sense to avoid the ever threatening Aardvark hole. The Tommies heard our advance and closed their ranks, raising their heads and stamping their feet threateningly before wheeling and making for the open plain in one well-knit body. Now the chase was well and truly on. The Thomson's gazelle is one of the fastest of all antelopes; few predators can catch him once he knows what he is up against. The cheetah can do it provided he can creep close enough before launching his final devastating charge, but then the cheetah is the acknowledged speed ace of the animal kingdom, a sort of four-footed Fangio. A pack of wild dogs, those relentless killers, can also do it, running down the victim in relays by sheer tireless endurance aided, I think, by the same sort of hypnotism that brings a full-grown unwounded rabbit eventually to the jaws of a hunting stoat, many times its inferior in speed and weight. Here, on its own ground against a single opponent, the gazelle's performance was likely to be very different.

All the group kept well together moving at three-quarter speed, while every few yards one or another member of the group would spring into the air in a skittish carefree sort of way, "pronking" as the Afrikaans call it when referring to their native springbok, a real master at the art. Tally, once he had got his quarry well and truly in his sight, wasted no time in such frivolity but sailed effortlessly over the short wiry grass, his plumed tail floating like a creamy banner behind him. After traveling together in a tight pack for some hundreds of yards one member of the group, a magnificent full-grown ram, probably the leader of the herd, broke away from the main body and swung right across Tally's front. Tally wheeled with him, cutting between him and the rest to make sure that he could not rejoin his companions who, quick to seize the opportunity, were rapidly disappearing in the distance, breasting a slope far out on our flank.

The old ram had given up his pronking and now settled down to some serious running; he was determined to show the solitary interloper what he could do in the way of record breaking. The ponies had entered into the spirit of the hunt and were storming over the sea of grass like a brace of destroyers while I was fully occupied keeping my seat and trying to guide Silver Wedding around, through and over the natural hazards as they presented themselves in kaleidoscopic variety, each trickier than the one before.

The buck, now well ahead, came to a narrow fissure, a long dried-up water course, raw and red against the grassy plain; without breaking his stride he took it and scudded onward. Tally, keeping his distance behind and neither gaining nor losing a yard though now fully extended, paused for a barely perceptible half second before he too gathered himself and cleared the gap like a Grand National winner. Grant shouted to me as he struggled to rein in his mount, whose mouth was as hard as the ground he was negotiating. I glanced toward the sound of his voice and saw the pair of them swing right and go pounding on, running parallel with the rift. I managed to get some sort of control over Silver Wedding just as she reached the edge of the crevasse, which was deep and narrow with a nasty take off; she shuddered, sat back on her haunches and plunged across. I was over, landing on her neck, but still in the saddle, and looked up in time to see hunter and hunted disappearing over a fold of ground. Silver Wedding seemed almost as surprised and relieved as I was to find herself safely on the further side of this rift valley; she steadied herself, shook her extended head and neck, gave what appeared to be a great sigh of satisfaction and thundered on after the buck and his pursuer, who had by now disappeared behind a tree-crowned rocky island that rose unexpectedly from the undulating sea of plain.

Grant had negotiated the *mullah*, and he and his mount

A Proud Saluki

were racing along the skyline at a hard gallop, but of the two principal performers there was no sign. Suddenly Grant let out a rousing "View Hallo," and by standing in my stirrups I could see a dark shape followed at perhaps a hundred yards interval by another smaller one. Both were coming our way: the leader moving with the tireless rhythm of an animated rocking horse; the second still following gamely, though steadily losing ground. As the buck streamed past, his dagger-sharp black horns appearing to rest on his withers, I noticed with surprise that he seemed to have grown in stature during the chase, until I realized that it was no longer a Thomson's gazelle that was leading the hunt but a fine male of the similar, but considerably larger, Grant's gazelle. Somewhere out of sight Tally had switched to a fresh quarry which was leading him, elusive as a will-o'-the-wisp, to the point of utter exhaustion.

He was a young dog and I feared he might overtax his strength and perhaps strain his heart in the fruitless effort to catch the virtually uncatchable. I rode in to cut him off, shouting his name; he appeared not to hear my voice but went determinedly past, his tongue out and his muzzle sprayed with froth. I shouted again and cantered after him; the buck was pulling away, and after a further two hundred yards or so Tally slowed down to a trot and finally drew up. His sides were heaving, and his long red tongue flickered in and out like a snake's as he turned reluctantly and trotted back to me. He was leering happily, his mouth wide open in a grin that made him look like a self-satisfied crocodile. How well I was to know that grin in the years that followed; his whole attitude seemed to say, "Just you wait till we meet again, I was only playing this time." In fact he had done very well; he had "broken the pronk," a feat few puppies of his age would have been capable of.

After a short rest we rode quietly on across the gently undulating plain, now bathed in the rosy glow of the rising sun. Tally recovered his wind in no time and had another short but exciting course after a Cape hare, which he managed to turn twice before it wisely if rather unsportingly went to ground in a convenient antbear hole. And so we turned and headed back to camp.

Reaching the summit of a low grassy plateau we saw below us what at first appeared to be a snow drift; the grass for a hundred or more square yards seemed to be buried under a pure white sheet. We rode slowly forward and stopped entranced at what we saw. What had at first looked like a heavy fall of snow proved on closer inspection to be a multitude of thousands, perhaps millions, of pure white butterflies, clinging to very bush and blade of grass tall enough to support them. We had come upon one of the great but seldom described phenomena of Africa, a resting horde of migratory butterflies, all apparently of the same species, very similar to the European small white, and all, it seemed, facing in the same direction, northward.

We dismounted a short distance from this immense snow-white, motionless yet living carpet, loosened our girths and let our ponies graze, while Tally flopped out on the comforting natural mattress of springy grass and began noisily to lick his paws. We lay on our backs and smoked much appreciated cigarettes; our horses munched contentedly nearby, and a yellow-throated long-claw warbled merrily, if none too tunefully, from the top of an isolated bush. The sun's growing warmth embraced and relaxed our tired, aching bodies; my muscles had already begun to stiffen with the strain of the sudden, unaccustomed exertion.

As we lay there, basking in the sunshine and enjoying a well-earned rest, a sudden message seemed to reach the sleeping white multitude close by, for the huge assembly began to vibrate with life. In their countless thousands the butterflies climbed deliberately up the stems of vegetation upon which they had perched throughout the dark hours and, clinging to the top, began to open and shut their wings, silently and rhythmically. The effect of this, as we sat there enraptured, was both exciting and dramatic, giving the appearance of a dazzling field of winged pyrethrum flowers swaying together in the diamond-bright highland air. Then, as if obeying orders of some secret leader, the butterflies rose a few feet into the air and, forming a great column, moved off in northerly direction, following the contours of the ground. Dipping and fluttering, but never settling, they passed slowly and resolutely onward, a great silvery-white river flowing over the plain until lost to view in the hazy shimmering distance. This was one of the most impressive and inspiring sights either of us had seen, and we spoke little as we walked and trotted the remaining few miles to the camp.

The next day we drove back to Manyani and shortly afterward Tally and I caught the evening train for Mombasa. Through the seemingly endless night I held him very close to me as the train chugged slowly along, winding its way through the starlit bush. I hated to part with him and yet I knew I had made the right decision. I had taken Tally on and now, come what may, he was my responsibility. I have nothing but contempt for the sort of dog owners who, at the slightest hint of inconvenience or expense, either have their dogs destroyed or, far worse, abandon them. The expense of boarding Tally at the quarantine kennels did not worry me unduly at the time. I hadn't the faintest idea how I was going to pay for it, but that contingency was six months and five thousand miles away.

When I reached Mombasa I hailed a taxi and was about to climb in when the driver, a strict and inflexible Moslem, started to cut up rough about taking Tally in his vehicle. I forbore to point out that to any self-respecting Arab of the true desert breed a Saluki was not a dog, but a hound of the blood royal, and as such entitled to enter even the tent of a Bedouin sheikh himself. I didn't bother to argue but I merely called another taxi, driven by a cheery African who apparently had no religion whatsoever. We drove to the quay, where the S.S. *Kenya Castle* rode at anchor, gleaming white against the intense blue of the Indian Ocean and the dusty gray-green of the few stunted palm trees that stood disconsolately in the background. She was a splendid vessel and my heart leaped at the sight of her. She was going home and my soul yearned to go with her.

I had had enough of Africa for the time being at least. I thought of the *Empire Ken* and of all the fun and enchantment of the outward voyage. It seemed a lifetime ago, though it was in fact eight years. Now I felt a sudden, almost overwhelming, longing for Sussex, "green Sussex fading into blue, with one gray glimpse of sea." Suddenly I could see it all so clearly; the great gentle sweep of the downs, the little villages nestling in their protective hollows, and I could almost hear the querulous insolent call of jackdaws at their nesting holes in crevices of the chalk cliffs that skirt the Channel at Rottingdean. Several passengers, their faces alight with excitement at the prospect of the voyage ahead and of their return to the Old Country, were hurrying up the gangway carrying zip bags, suitcases and folded copies of the *East African Times*.

I took a firm hold of Tally's lead, patted his head, and strode up the gangplank after the passengers. I noticed curious glances cast in our direction; perhaps Tally was being mistaken for a police dog in disguise and it was imagined we were there to rout out some felons from their lurking place in the bowels of the ship. As I reached the deck and paused wondering what to do next a large figure detached itself from a small

A Proud Saluki

knot of seamen and came forward with a friendly grin. He had enormous hands, and was clad in what looked like a boiler suit or dungarees. "Well, sir," he said, "we have been expecting you and your pal. What a fine dog he is! We'll take good care of him on the voyage." This unexpected ministering angel proved to be the ship's cook, the traditional custodian of such hostages to fortune.

He made a fuss of Tally, who accepted it as his due but without any great show of enthusiasm, a trait of Saluki nature that, while flattering to those lucky enough to enjoy their affection and respect, can be decidedly embarrassing for those who do not. No Saluki bothers to conceal its feelings and, if bored or displeased by uninvited attention, makes no effort to hide the fact, which is to my way of thinking a decided point in its favor. The cook, a sensible and understanding fellow, asked me to follow him which I did, trying to fight off the icy hollow feeling that seemed to pierce right through me. I had made up my mind, had paid for the passage and there was nothing I could do now except hand Tally over to his keeper and clear out, as quickly as possible.

As I threaded my way in the wake of the cook, who was trying to cheer me up, another figure was watching us intently. She rose from a deck chair, where she had been lounging, came toward us and, putting her arms round Tally's neck, stroked his long silky black ears. Tally for his part treated her in the same embarrassingly cavalier way that he had treated the cook. The girl at last turned and spoke to me. I am not unduly susceptible to feminine pulchritude, unless the charms consist of a hooked beak and talons on the toes, or alternatively a furry coat and long furry tail; but I felt no such prejudices now. Tall and slight, with honey-blond hair, she had eyes of the most extraordinary clear amber. Her name, Griselda Trelawney, was as unusual as her appearance, and she was on her way to England from her home in Durban. Having been seriously ill with some obscure fever she was going to stay with relatives for a few months, hoping that a change of climate and scenery would make her well again. She had left behind an Afghan hound, her only pet and constant companion. The dog was being looked after by her parents, who loved it almost as much as she did. The sudden appearance of Tally had reawakened her longing and nostalgia; she showed me a photo of the Afghan, and I knew exactly how she felt. Akbar was the exact color of Tally and had the same air of gentle aloofness, which is so much a part of the make-up of these two closely related breeds.

We came at last to a sheltered part of the deck where, under an awning, stood a row of solidly made kennels. Tally was not to be the only dog on the ship, for peering forlornly from one of the kennels was a year-old boxer puppy. With his black face and huge expressive eyes he looked like one of the old-fashioned black minstrels that used to be so much a part of the summer scene in English seaside resorts. His owner obviously thought so too, because he had given him a singularly appropriate name, Rastus. Rastus had boarded the ship at Dar-es-Salaam and was on his way to six months incarceration in the well-known quarantine kennels at Hackbridge. He pushed a huge paw against the wire grille of his kennel, desperately seeking human contact and reassurance. I stroked it with two fingers, and like to think he felt happier. Tally ignored him utterly, though I heard later that they became friendly, seeking solace in each other's company.

I patted Tally, holding his head as he pushed his long muzzle into my armpit in the disarmingly intimate way he had. I helped slip him into his kennel and hurried away. The cook assured me that each dog was given daily exercise on a quiet part of the deck, and had the best possible attention. I thanked him and left. I could see no point in lingering there, the break had to come and to delay any longer only made things worse.

A Proud Saluki

Griselda walked with me to the gangway, telling me that she would personally keep an eye on Tally, would watch his disembarkation and would write and let me know how he got on. When I reached the bottom of the gangway, I turned to wave and was swallowed up in the throng of brightly clad Africans, Arabs, Indians and all the others that make up the population of a port such as Mombasa.

13

Kwaheri Kenya

NEEDING A DRINK, I headed swiftly for the best hotel. Turning a corner I nearly fell over Fergus McBain. Seldom have I been more pleased to meet anyone. Fergus, still in the Locust Control, was on his way back to Voi after one of his frequent surreptitious trips to Malindi. We drove out to the Nyali Beach Hotel where, deep in comfortable chairs, with iced drinks at our elbows, we talked about all that had happened since we had last met. Of how Fergus, coming suddenly upon a lioness with cubs somewhere near Carissa, had been attacked and how only deft use of the steering wheel on his land-rover and powers of acceleration had saved his life.

As we drank and chatted I began to relax, although my heart and thoughts were still aboard the *Kenya Castle*, now just about to begin her long haul northward toward the horn of Africa and the Gulf of Aden. Fergus offered me a lift back to Manyani, but as he had to call on someone in Moshi it would be a somewhat Chestertonian journey through a considerable area of Tanganyika. I accepted all the same. I was still officially on the staff of the camp at Manyani but nobody seemed to be unduly worried as to whether I was present or

Kwaheri Kenya

not, and my replacement had already arrived to take over my duties. I was confident, too, that Grant would look after Karen efficiently until my return.

We drove off at length traveling almost due westward. At Voi, Fergus paused to drop off some of his belongings and to write up his official diary and log book. Then we took the road to Moshi, some considerable distance across the Tanganyika frontier. Fergus drove with his usual panache and we spent the time discussing the works of Henry Williamson, on which he was a considerable authority. At Moshi we had a pleasant night in the hotel where years before I had begun to realize that I was at last in Africa and that my new life was about to begin.

After a leisurely breakfast, and while Fergus went in search of petrol, I explored the tangled garden I remembered so well. The sunbirds were still busy among the scarlet poinsettia flowers. I shall always have a special feeling for sunbirds; after all they were the first truly tropical birds I had ever seen in their own homeland. We were soon racing in the direction of Voi once more. This time we did not speak much; my mind was full of plans and I did not have too much time to lose.

I might have finished with Africa but Africa it seemed had not yet quite finished with me. Somewhere between Moshi and Voi we nearly ran over what looked like a refugee from the cast of the film *The Lost World* slowly crossing from one forest-skirted side of the road to the other, steadily and deliberately as if it had all the time in the world. It was a scaled-down pre-historic reptile complete to the last detail. We stopped the land-rover and ran back to inspect this intriguing creature that turned out to be a chameleon, being at least eighteen inches from the peak of its archaic head to the end of its prehensile tail.

The thought occurred to us both almost simultaneously that he might be thirsty. We knew that chameleons did not drink in the conventional way by lapping water from a pool or stream, or even from a water-filled hollow among the branches; individualists as they are, they prefer to obtain their moisture by "shooting" drops of dew or rainwater with their versatile tongues. Fergus broke off a spray of the fleshy silvery-green leaves of a euphorbia tree, and we sprinkled it with drops of tepid water from our drinking bottle; sure enough our new traveling companion eyed the water-beaded leaf for a moment fixedly, swaying backward and forward on bent legs as if taking aim, shot out what seemed like half a yard of tongue, transfixed a globule of moisture and withdrew tongue and water drop in one lightning movement. He looked at me with what appeared to be a grin of satisfaction and took aim once more; repeating this performance several times until his thirst had been quenched, whereupon he swiveled his eyes around, each one moving independently of the other, took a firm hold on the branch and settled down to ruminate in his own detached philosophical manner.

Thus, still grasping his chosen perch, he accompanied us on our journey to Manyani where, as I had expected, he cause a fair amount of consternation. The Africans are, or were in those days at least, terrified of these harmless and attractive reptiles, endowing them with all manner of deviltry and unspeakable *ju-ju*. I always found this odd, as the Africans must have encountered them frequently in their journeyings to and fro in the bush, and one would have supposed that in time they might have realized that, for all their sinister appearance, they are utterly without malice and are in fact extremely useful beasts doing their fair share of insect destruction. But the arrival of this splendid specimen caused near panic among the askaris and even some of the Europeans were a bit dubious as to just how innocuous he was.

I fastened his branch to the ceiling of my room by a short length of string and here he was perfectly at home although I found that if he wanted to do a bit of exploring he would just let go and plop onto the floor beneath. This seemed to do him no harm; but having repeated his falling act three times in succession I decided to let him wander where his fancy led him, which happened to be my pillow upon which my head was resting at the time. For lack of mice and other small rodents, which I later found him well able to tackle, I gave him large grasshoppers and an assortment of the huge armor-plated beetles that always appear from nowhere as soon as a light is turned on. In his own enigmatic way he was a considerable personality; his very size was impressive, and the timeless look of world-weariness with which he viewed his surroundings sometimes made me feel that he must have been around since the beginning of creation. This somewhat cynical expression reminded me of a dubious character who, mistaking me for an American officer, tried to sell me the Zoological Gardens in Regents Park when I was on a visit to that time-honored institution shortly before my departure for Africa. I have no idea what his name was, but for some reason I have always thought of him as Sid; anyway the name suited my Morogoro chameleon.

My time in Kenya was running out and I wrote to all the airline companies that included Nairobi in their itinerary to book passage to the U.K. The Comet had made its maiden flight to and from South Africa some time previously, and long distance jet aircraft were becoming increasingly commonplace; however I mistrusted these pressurized supersonic monsters. Besides, they were far too expensive for me. Finally I managed to get a seat on a Viking belonging to a firm called "Hunting Air Transport." The Viking was a nice old-fashioned propeller airplane that took a leisurely two and a half days for the trip, which suited me admirably. I did not consider omitting that I would be accompanied by a falcon and a giant chameleon a particularly heinous piece of villainy, for

I knew that it was not illegal to import birds or reptiles into Britain, and that they were not (with the exception of parrots) subject to quarantine regulations. But I also knew that I might be liable for duty that I couldn't afford and, far worse, that the animals might not be allowed to travel on a passenger plane; or if they were they might find themselves going the equivalent of aerial steerage. What I proposed doing was, I considered, no more desperate a crime than the concealment of one bottle of hooch over the permitted limit. In any case it added a certain spice to the undertaking.

The last few weeks sped by as they are wont to do when big happenings are afoot. I got a letter from Griselda Trelawney telling me that Tally had arrived safely and had settled down in his comfortable if somewhat circumscribed place of temporary incarceration. He was eating well, appeared to have accepted his lot and had made friends with his own particular, very understanding lady "jailer" who was doing all she could to make him as happy as possible. This cheered me up considerably! I knew that if he was feeling well there was no need for concern. After all I had survived two and a half years in confinement, so he should be able to stand six months.

In no time I found myself on my last long trip in Africa, driving up the Mombasa to Nairobi road with Grant at the wheel, Karen on my fist and Sid the chameleon, self-contained as ever, grasping a large branch that rested athwart my two bulging suitcases at the back of the truck. As the battered vehicle jolted and jumped westward I looked out at the distant blue horizons I had known and loved for eight long years and that had become a part of my very soul. I saw one of my old friends, a chanting goshawk, perched on the summit of an acacia by the roadside and as we passed he peered straight at us for a second, unafraid and at home. High overhead swung the delta-winged shape of a Bateleur, perhaps the most characteristic of all African bird life; seeing him racing high above us I thought of Torquil and wondered how he was making out. Far to the southwest of Kilimanjaro lay unseen the forested heights of Oldeani where rested Cressida and Rupert, two of the best friends I had ever had.

I turned my thoughts away and, as Karen shifted her grip on my fist and turned to look at me with her lovely dark eyes, the thought occurred to me, could she be the reincarnation of Cressida? They had much in common both in looks and mannerisms; however unlikely this might be, the idea gave me needed cheering up, for much as I longed to get home and to see once more familiar places and people I was sorry to be leaving Africa. The country had been kind to me and I was going to miss it deeply.

Near Machakos we paused to buy some Coca-Cola from a wayside *duka* and to drink it at the side of the road. As we lay in the long grass gazing upward, a party of European swallows came swinging past low above the grassland; they were conversing musically together in the intimate confidential way swallows have. It was early April and they were going home to Europe to nest and rear their families, perhaps even to the very loft at Old Acres where, ever since I could remember, a pair of swallows had their summertime headquarters. I would be following hard in their wake and my best wishes went with them as they scudded happily northward following their age-old skyway to the place of their origin.

In Nairobi I booked into the Queen's Hotel. I was very busy with all the inevitable and frustrating trivialities that dog the footsteps of the home-going expatriate. Passports had to be brought up-to-date, currency changed, visas arranged for landing in non-British and somewhat unfriendly territory (in this case Egypt), and a great many other things besides. I managed to call on a number of old and new acquaintances and paid a final and rather nostalgic visit to the Coryndon Museum where I had a good look at their specimen of the

Morogoro chameleon. I noted, with considerable satisfaction, that it was completely dwarfed by Sid.

With no little ingenuity I made a sort of pouch and fitted it into the lining of my sports jacket, and after a bit of preliminary reluctance I persuaded Karen to repose therein. I was relieved to see, when inspecting myself in a mirror, that no suspicious bulge appeared and under an overcoat it would be infinitesimal. This was highly satisfactory for I had been none too certain how I was to overcome the problem of the prying eye of officialdom. Luckily Karen herself was cooperative, and once at rest in the warm darkness she became as tranquil as if sitting on her nest and brooding her young.

In the case of Sid I just hoped for the best. I bored some ventilation holes as inconspicuously as possible in the smaller of my two suitcases, fixed a substantial twiggy perch from corner to corner, ensconced him upon it, tickled his chin, bade him *bon voyage* and hoped that he would be comfortable. Feeding him would present no problems as chameleons do not eat very often, and a couple of days without food would inconvenience Sid, who was almost indecently fat, not one whit. All my other possessions, such as they were, were crammed into the other case.

On the evening before I was due to leave I took Karen to the highest point of the escarpment, where the Nairobi road begins its dramatic drop down to Elmenteita and Navasha. This was almost certainly the last time that I would see Karen in flight against the backdrop of the wonderful Riff Valley, in the heart of the country where she and I had so fortuitously been brought together. As she wheeled high overhead a small party of wire-tailed swallows began to mob her and she quickly responded, dropping with half-open wings straight through the chattering flock that melted away at her coming. The sight of her, dusky as night save for her glistening white cheek patches, hurtling along the terraced hillside, was a memory I shall always treasure. Away to the southwest rose the craggy crater of Mount Longanot's extinct volcano, a great natural castle, thrusting upward out of the far-distant horizon. This to me was Africa indeed! I called Karen down after she had put on a superlative performance and together we drove back to the coffee estate at Kianbu where I was to spend my last evening in Kenya.

Dawn found me packed and waiting for the airport bus. Karen reclined sleepily and happily in her hawk container and Sid, bulging with a special feed of succulent locusts and praying mantis, goggled at me as I made sure he was comfortable on his perch in the otherwise innocent-looking suitcase. It was raining when we reached the airport just before dawn, as indeed it always seemed to be whenever I went anywhere of import. The plane, all I could see of it, looked workmanlike and reliable enough. My only previous experiences of long distance air travel had been as a guest of the Third Reich, when I had made an unpleasant journey from Sicily to Italy in a JU52 with a lot of equally frightened and equally dispirited POWs, and later as a guest of the American Air Force when I was flown home to Britain in 1945. In that case the plane had been a Dakota, but it could well have been a chariot of fire so elated was I at the prospect of freedom ahead.

Among my fellow travelers were several whom I knew, but they were all married couples with or without families. It seemed that I was the only unattached male and as such found myself allotted a seat right up in the bow, just behind the crew's quarters. The two seats immediately opposite me were occupied by very attractive girls and there was another, equally decorative, beside me next to the window. The air hostess herself was as nubile as one might have hoped, a Latin-American named Dolores. The trip augured well if this was a sample of what lay ahead. However, the lavatory, I noticed with concern, was in the stern. I knew that a considerable amount of my time would be spent there exercising and feeding Karen, who up to now had been behaving in an exemplary fashion and keeping tactfully quiescent.

Soon we had fastened our seat belts and listened to a welcoming address by the Captain; the propellers began to whirl, the plane to throb and vibrate with sudden life, and soon she was charging down the runway and had lifted gently into the air with scarce a sign to indicate that she was airborne. We headed out into the murky half-light of a new day passing high over Longanot and Hell's Gate Pass with its colony of griffon vultures wheeling below, over Nakuru and away to the northwest at a (by modern standards) leisurely two hundred fifty miles per hour. First stop was at Kampala Airport in Uganda, which was as hot, dry and sunburned as one might have expected. On the roof of one of the buildings, like a deputation come to speed me on my way, sat a croaking row of pied crows, which appropriately were the last as well as the first African birds I was to see. Shortly after leaving Kampala I paid my first visit to the lavatory to feed Karen and to let her stretch her wings and preen her feathers, which as always were immaculate. Unperturbed she ate largely, but understandably was none too keen to be returned to her dark seclusion; so I felt beholden to extract and exercise her more frequently. After my fourth visit I noticed the girls eyeing me with sympathy and the hostess, who for some time had been watching me and my yo-yo-like progresses, asked tactfully if I was feeling all right. It was a somewhat embarrassing situation, but I managed to jolly myself out of it.

The first night stop was at Wadi Halfa in the Sudan; my only impression was that the heat was more intense than I could remember anywhere, even in the Northern Frontier Division of Kenya. I spent a pleasant evening drinking beer with the crew, whose quarters I shared as all the official accommodation had been taken over by the families. In a moment of unusual expansiveness I decided to share my secret (at least part of it—I did not mention Sid). I produced Karen, who immediately carried the day, although I was told that I was very naughty indeed, and that on no account must I tell the other passengers that officially and regrettably Karen must still be considered *persona non grata*. However Dolores, with the ingenuity that one expects from someone in her profession, conjured up a piece of fresh goat's meat, which was particularly welcome as the meat I had so carefully stowed away before departure was looking and smelling a bit unsavory.

In the quiet seclusion of the sleeping annex I perched Karen upon the bedrail and extracted Sid from his suitcase so that I could have a good look at him. He had lived up to his reputation, having changed color to an almost inky blackness, so different from his usual grassy green as to make him appear a different animal. His best effort prior to this had been to turn a rather sickly pink when placed for an airing upon a pile of bricks back at the Camp at Manyani. The journey so far did not appear to have upset him although he was thirsty and enjoyed "shooting" drops of water sprinkled about his perch. I was too excited to sleep much that night. The mosquitoes were, if possible, more numerous and voracious than any I had previously encountered in Africa and my mosquito net had a hole the size of my head in it through which the loathsome legions trooped in endless succession in their search for good British blood.

The next morning we took off again, after many cups of tea. Breakfast I could hardly look at; fried eggs and bacon at 6 A.M. in a temperature of ninety degrees or worse is not my idea of an ideal meal, but like fish and chips they seem to be accepted as part of the British way of life. I took care to give Karen as much exercise as I could in the short time available, flying her to the fist a number of times across the sparsely furnished room in which I had slept, and once more she settled down in her traveling place without any sign of distress. By now the passengers had got to know each other a bit and the British reserve had begun to thaw out so the whole atmosphere was more relaxed and cheerful. One of the young women in the seat opposite me had a baby girl in a carry cot. The baby was only about three months old and slept almost continually throughout the flight; I must confess I envied her.

That night we were due to arrive in Malta for our second and final stop. At first the aircraft droned slowly and monotonously onward. I had noticed on first entering the airplane at Nairobi that each seat had a slot attached to it and that each slot contained a large brown paper bag. I had wondered idly what these paper bags were for. I was only too soon to find out. As the Viking roared northward over the desert, which we could see stretching away like a great yellowish sheet beneath us, the sun began to climb and hot currents of air to rise. Without the slightest warning the aircraft appeared suddenly to fall out of the sky as if seized by some gigantic hand thrust upward from the ground; how far she fell I do not know, but within seconds she was shooting upward again, only to fall once more. I thought we had had it! As I glanced wildly around I could see that I was not alone in the belief. Dolores, however, seemed unperturbed, which cheered me up considerably.

The Viking continued to be used as a plaything by the elements and soon everyone was too occupied with their paper bags to worry about the prospect of imminent demise. Personally, I didn't care much what happened as long as it happened quickly and painlessly. I made a vow there and then that, if fate decreed that I should cross the Sahara again, which heaven forbid, I would do it in safety and comfort by camel. Things then calmed down a bit, though we had a further mild buffeting before we landed limp, languid but alive at Cairo Airport for a refuel. The surly looks of the airport officials did little to spoil our pleasure at being once more on ground level with something a bit more solid than the floor of our airplane beneath our feet.

After a short stop we flew on above the choppy white-flecked Mediterranean to touch down at Valetta. I had never been to Malta and was intrigued by all I saw. We stayed at the Phoenicia Hotel, which was the most luxurious establishment I had encountered in years. Dolores, knowing my predicament and with her usual tact and ability to pull strings, arranged for me to have a single room where I could liberate Karen and let Sid stretch his legs and climb about on the curtains. I found two or three enormous crickets in the garden of the hotel and he tongued them down with none of his usual rather theatrical preliminaries.

Early next morning we were off again, passing in brilliant sunshine high over the Alps. The tiny gray ribbon of the English Channel was behind us almost before we came to it, and it was late afternoon when we touched down at Heathrow Airport, which was hidden beneath a blanket of driving rain. I was not looking forward to the next stage: confrontation with Customs. I heaved my two cases onto the counter in as offhand a manner as I could manage and I was given the usual card to read which enumerated all the items that were dutiable or just plain prohibited. I had no firearms, new cameras, drugs or bottles of liquor, and I told him so. The gold-ringed inquisitor on the other side of the barrier looked at me coldly and tapped one of the suitcases. "Would you please open this, sir," he asked dispassionately. With semiparalyzed fingers I did my best to obey. Fumbling clumsily with the lock I took a deep breath and flung open the lid. There, exposed to the merciless eye of authority, crouched Sid, in all his prehistoric splendor, his eyes swiveling and his helmeted head raised truculently.

The Customs Officer was no doubt a brave man and must have seen many strange things in his time, but he had never met anything like Sid. After the initial shock he summoned a senior colleague. "What in heaven's name is this?" he asked, keeping well behind the barrier. "This," I replied, wishing to get the ordeal over as quickly as possible, "is a giant Morogoro chameleon, very powerful *ju-ju*, saved my life from the Mau Mau, wouldn't travel anywhere without it." To my astonishment and relief he indicated that I might close the suitcase; he then put a chalk cross on it and without further words pointed to the exit. I saw my mother among the crowd outside, her face one big welcoming smile. We greeted each other in the rather self-conscious way that results from reunions in the public gaze, hurried out to the waiting car and drove straight to my grandmother's house in Eccleston Street. Here I introduced her with some misgiving to Karen and Sid, both of which, as usual, she accepted with equanimity.

The next day, somewhat bemused by the crowds and the noise, I took Karen to the Natural History Musem. I traveled by underground and Karen, who up to then had never seen more than half a dozen or so people at one time, never batted an eyelid. At the museum, Derek Goodwin, the curator of birds, quickly confirmed what I had long suspected; Karen was indeed an Eleanora's falcon, possibly the first to arrive in Britain. The same afternoon I went to the quarantine kennels at Woolwich; it was with a mixture of excitement and trepidation that I entered the reception block and met the manager. He introduced me to Diana, Tally's own particular kennel maid, who was charming, efficient and kindly, yet without a trace of sentimentality. I knew at once that my dog would be all right with her.

I followed her down a long tiled corridor, and outside to a range of large well-built roomy kennels. Tally was standing by the door of his kennel, his head raised and his plumed tail waving gently. I spoke to him, his ears flickered and a sudden warm light of recognition came into his eyes. He stood on his

Kwaheri Kenya

hind legs and reared up against the barred door. Diana opened it, clipped a lead to his collar and led him outside. He looked magnificent, his coat gleamed and his eyes sparkled with pleasure; he knew me without a second's hesitation. I noticed that he had grown considerably in the two and a half months since we had been together. He was taller, better proportioned, had filled out and was looking altogether much more mature. We took him to the larger exercising enclosure and let him loose. Here he did his best to show how pleased he was to see me again, leaping and cavorting, but it was a pitiful travesty of the real thing.

As I watched I called to mind the great rolling ocean of grass, the lone acacia with its sentinel augur buzzard perched aloft, and the tireless surging gallop of the hunting Saluki. I vowed there and then that I would come no more to the kennels until the day of release. Back in his quarters I fondled him, gave him a huge meaty bone that I had brought as a parting present, and stole quietly away, not daring to look into the hurt, puzzled eyes that I knew were watching my retreating back. He had three months of his sentence still to serve.

The long hot summer passed serenely by and with it passed Karen. I had been flying her high up on the Downs above Lewes, and she had reached a pitch such as she had never attained before. She was cruising in great circles; her mind obviously elsewhere. She ignored the lure and continued cruising as if seeking something. Suddenly she veered and shot away southeastward down the path of the wind, cut over the Channel until lost to view in the shimmering distance where sea and sky meet. I watched half-hopefully, swinging the lure idly, but I knew that she had gone forever. I must have known deep within me that this would happen eventually, because only a few days before I had cut short her jesses, thus averting the danger of her getting entangled should she decide to strike out on her own. It was with a mixture of sadness and relief that I returned home hawkless; I knew that Karen could cope with life and make her way in the world.

I had much to occupy my mind. In a few days Tally would be coming home and I was going to make sure that he had a welcome worthy of him; nothing was too much trouble. I went to the best shop I could think of and bought a handsome collar and lead and, not being too happy about the British autumn which was close upon us, I also bought a splendid coat, worthy I thought for one of such ancient and noble lineage. That evening, as I sat thinking of the day of his release, the telephone rang, breaking my reverie. I picked up the receiver, and listened to the voice of destiny. "Would you," it asked, "like to have a young golden eagle?" I replied with as much conviction as I could raise: "I am sorry, I don't want any eagle, golden or otherwise." "But," continued the voice, "I just returned from Spain with two eaglets I rescued from a gamekeeper. As I'm living in a small flat in London I can't keep them myself. I've found a home for one of them, the male; would you at least have a look at the other one?" I weakened —at least there could be no harm in just looking at her. She came; I saw and I was conquered. Life has never been the same since. She came at random—and Random she has remained; but that, as they say, is another story.